Owens

FUTURE WORLDS

FUTURE WORLDS

John Gribbin
for the Science Policy Research Unit

University of Sussex
Brighton, England

PLENUM PRESS • NEW YORK AND LONDON

Library of Congress Cataloging in Publication Data

Gribbin, John R.
  Future worlds.

  Originally published: London: Abacus, 1979.
  Includes bibliographical references and index.
  1. Forecasting. 2. Twenty-first century—Forecasts. I. University of Sussex.
Science Policy Research Unit. II. Title.
CB158.G74 1981   303.4   81-5135
ISBN 0-306-40780-9   AACR2

1981 First Plenum Printing
©1979 The Science Policy Research Unit, University of Sussex, and John Gribbin

Plenum Press is a Division of Plenum Publishing Corporation
233 Spring Street, New York, N.Y. 10013

Printed in the United States of America

For
Chris Freeman
and
Marie Jahoda
who gave me the opportunity
to write it!

Contents

Simple diet is best, for many dishes
bring many diseases, and rich sauces
are worse than even heaping
several meats upon each other.

Pliny

Introduction

During the middle and late 1960s, concern about the way the world might be going began to move out of the arena of academic debate amongst specialists, and became a topic of almost everyday interest to millions of people. Concern about mankind's disruption of the natural balance of 'the environment' brought the term 'ecology' into widespread use, though not always with the meaning to be found in the dictionary, and fears that world population might be growing so rapidly that very soon we would run out of food, resulting in mass starvation and a disastrous collapse of civilisation, helped to make books such as *The Limits to Growth* bestsellers in the early 1970s. Today, quite rightly, decisions on long-term policy with widespread repercussions – most notably, those concerning nuclear energy planning – are a subject of equally widespread public discussion. But all too often such debate focuses on specific issues without the problems ever being related effectively to an overall vision of where the world is going and how it is going to get there.

At the Science Policy Research Unit, University of Sussex, a group working on studies of social and technological alternatives for the future has been contributing to 'the futures debate' for several years, cautiously (perhaps, in a sense, almost too cautiously!) developing a secure foundation for forecasting the way the world may develop. So far, this team has been best known for its attempts to counter some of the wilder forecasts of the doomsday brigade, first in a specific critique of *The Limits to Growth, Thinking About the Future* (US title, *Models of Doom*) and then in the more general methodological

9

work published in 1975 as *The Art of Anticipation*. With this scene setting complete, the team moved into the realm of forecasting proper and has now published a major volume, *World Futures*, investigating the ways in which the future of humankind on Earth may develop over the next fifty years or so.

Clearly, this landmark in the development of the futures studies programme at SPRU required that the resulting vision of the 'best' alternative for the future should be presented to as wide an audience as possible; equally, however, the responsibility of the team to explain why one of several possibilities emerges as the 'best', and to justify their arguments at a level appropriate to the 'futures debate' among specialists had overriding initial priority. This dilemma has been resolved by producing two books setting out the SPRU vision of the future. *World Futures* is a necessarily large volume setting this work in its proper context and looking in detail at four possible ways in which the world might develop from the situation which exists as we leave the 1970s and enter the 1980s. While the authors of that book have creditably avoided the kind of specialised 'erudition' which is meaningless to all but a few specialists, the sheer weight of background material, and the nitty-gritty, detailed points which must be covered, make such a work inevitably intimidating to the more casual reader – although such readers are just the people who will be affected by the way the world goes over the next half century.

So the present book was conceived as a counterpart to *World Futures*, setting out more compactly the main theme of our hoped-for 'best' future world, a future in which economic and industrial growth can be combined with a more equitable distribution of the wealth of the world to ensure reasonable standards for all. Such a possibility sounds so much like head in the clouds woolly idealism, compared with the doomsday dramas propounded by others, that it would be crazy to have put the present account forward first, without the sound basis of

10

World Futures to back the vision up with solid reality. But the present book is in no sense a 'condensed version' of that companion volume. Both draw on the same body of work, carried out at SPRU, and both tell the same story of the 'best' future world we are likely to live in, if we make the right decisions now. The present book, however, not only leaves out of consideration some of the material used in *World Futures*, but also covers some ground ignored in that book. And, of course, whereas *World Futures* is in some sense an official statement of the SPRU forecasting team, *Future Worlds* is essentially my own interpretation of the work of that team, presented with a flavouring of my own views and written more because I felt that here is a story worth telling than because of any commitment in terms of academic standing and career.

This approach has been made possible because of my position as a contributor to the work of the forecasting team since 1975, but as an outsider not an official team member. My own area of specialist interest was, and is, the way changes in climate over the next fifty to one hundred years are likely to affect society, but in order to understand that you need to know not only how climate is likely to change (difficult enough in itself!) but also how society is likely to change in other ways, quite apart from climatic pressures. I have been with the team for exactly the period during which these ideas were being developed, and write here primarily as a reporter given a rare opportunity to observe the researchers at work. While this book does, overall, have the seal of approval of SPRU as a fair report of that work, the members of that research team should not be held responsible for whatever sins of omission and glib simplification I have committed in the reporting process.

That said, what is the basis of the story I have to tell? In a nutshell, the essential difference between the futures debate today and the situation ten or so years ago is the more widespread realisation that it is nonsense to talk about 'the' future and to describe what 'will' happen. Rather, we can see clearly that by starting out from the situation today we can arrive at any of several possible future worlds, depending on the decisions we make deliberately, or the decisions we allow to go by default. Our vision is not of what 'will' happen, but of what *may* happen in certain circumstances. As a cautionary tale to make the point clear, we need look no further than a statement made in 1969 by U Thant, then Secretary-General of the United Nations, and quoted in *The Limits to Growth*:

> I do not wish to seem overdramatic, but I can only conclude from the information that is available to me as Secretary-General, that the Members of the United Nations have perhaps ten years left in which to subordinate their ancient quarrels and launch a global partnership to curb the arms race, to improve the human environment, to defuse the population explosion, and to supply the required momentum to development efforts. If such a global partnership is not forged within the next decade, then I very much fear that the problems I have mentioned will have reached such staggering proportions that they will be beyond our capacity to control.

Well, here we are a decade later, and the problems are still not under control. But, equally, they still don't look beyond our capacity to control, and even the most concerned voices can be heard muttering to the effect that 'if we don't do something dramatic within ten years it will be too late'. The lesson is not so much that U Thant, or the mutterers of today, can be proved 'wrong' but rather that they have not looked at alternatives. The world system *may* run out of steam in the near future, or it may not; the worst aspect of doomsday scenarios is that by their apparent inevitability they encourage a *laissez-faire* attitude in which people live it up while they can, for tomorrow may never come. The sober reality that tomorrow may come, if we pull in the right direction, is in some ways a less welcome touch of cold reality.

12

The whole business of futures studies is likened by its critics to science fiction – but as a science-fiction reader I interpret this criticism as praise, and indeed we can draw an analogy with the position human society is in today with one of the ideas beloved of science-fiction writers, the possible existence of alternative (or parallel) universes. These take many forms, starting out from the premise 'what if . . .' What if, in a historical context, the Spanish Armada had succeeded in conquering England? What if the Norse had not just discovered America but established permanent settlements? And so on. At every point in time, the argument runs, a choice of possibilities lies ahead, an infinite variety of parallel worlds.*

Our visions of the future are, in a sense, equivalent to this concept of possible parallel worlds. We can select various possibilities from the array of future worlds before us, and by choosing an appropriate course of action decide which of those worlds we will inhabit. Some future worlds are more likely to develop than others, given our limited ability to change the way things are going; some are more desirable than others, among the choice of plausible possibilities. It is this question of desirability that has provided a focus for the SPRU futures work, and *the* focus for the present book.

First, let's rule out the wild extremes of implausibility, the real science-fiction material on the lunatic fringe of the futures debate. At the most unpleasant extreme, various prospects of natural disaster such as a new Ice Age setting in overnight, or the Earth being torn apart by earthquakes and drowned in a new flood of global proportions, have been presented by more or less wild-eyed visionaries. At the extreme of comforting reassurance in the future of mankind, pressing ever onward and upward to greater and more widespread glory, lie the visions of an inter-planetary – even interstellar – future for human society. The prospect of a truly epic global natural disaster must

* This idea, believe it or not, is also touted as a serious scientific possibility, developing from quantum theory. But we can leave that debate to the philosophers!

be ignored here, since the effects on society would be so great that development afterwards would follow a completely different path, bearing little or no relation to the future worlds that can develop from the present situation in the absence of such disaster. Such a prospect is also, happily, extremely remote: the Earth has provided much the same environment for life over the past sixty-five million years, and even ice ages don't count as truly global disasters. There is a hint of some catastrophe – perhaps a reversal of the Earth's magnetic field - which helped to bring about the end of the age of the dinosaurs about sixty-five million years ago. But such changes, short-term in the long history of our planet, do give rather more warning of their coming than the fifty years or so we are concerned with here.

It is this question of timescales, too, which rules out of consideration the spacefaring visions. O'Neill and his disciples are probably right in saying that we now have the technological capability to develop 'habitats' in Earth orbit (see *The High Frontier*), but there is no reason to see such a development being given a high priority over the next five decades, at least in terms of shifting a noticeable fraction of the Earth's population into cities in space. The vast majority of the world's population will still be living on the Earth's surface for our lifetimes, so it is natural to focus our attention on the future of mankind at the Earth's surface.

This does not, however, mean that there will be no significant developments in the exploitation of near space, and the more low-key visions flowing from O'Neill's work – factories in Earth orbit, power satellites beaming solar energy to the ground, and so on – are among the 'wild cards' which will ensure that the detail of any future world will inevitably be different from anything we can imagine today. Perhaps this is just as well – we must leave something for the next generation of futurologists to work on!

Within the framework of plausibility, then, our starting

14

point for developing visions of the possible future worlds must take considerable account of the problems now facing mankind – the problems which are themselves the basic cause of the concern about the future which has produced the widespread debate, ranging from the sublime to the ridiculous, of the past fifteen years or so.

There are three main features of these basic problems which we have to take account of. The most obvious feature, the underlying theme of the entire futures debate, is the dramatic rate of economic growth experienced by the developed countries of the world since the Second World War. Can this growth continue? Is it desirable that it should continue, even if possible? Or does ever-increasing individual consumption, the fuel which has driven this spectacular growth, inevitably imply a waste of resources and destruction of the environment?

So our three points for discussion are the possibility of growth, the desirability of growth and, by no means least, the effects of this growth on the less-developed countries, the poor relations who now, and increasingly, are naturally keen to share in the riches produced by this growth. The last point is at the heart of the SPRU approach to finding a pleasant future world in which to live; the subjective standpoint of all of this work is that of moving towards a reduction in these inequalities between rich and poor, and this inevitably colours the whole of the forecasting work. But such a subjective choice is not, of course, purely altruistic. If the one real prospect of destroying the entire basis of all this forecasting is the possibility of widespread war (see Chapter Two), then it is clear that one cause of such war could be the existence of unacceptable inequalities between rich and poor. By sharing the cake more evenly we not only sooth our consciences, but also take out insurance against violent conflict.

The widening gap between rich and poor is something which worries many people today, not just the futurologists in their academic havens. During the post-war period of rapid growth, the gap in terms of income per head

15

between industrial nations and the poorest countries has grown so obviously as to cause widespread concern – although it's not always clear just which gaps we ought to be worrying about.

In terms of improving mortality and literacy in the poorest countries, for example, there is a much closer resemblance to the situation in the rich nations. The economists who argue, however, that the existence of an 'economic' gap (in terms of income per head) is of little importance compared with the provision of social welfare in the poor countries clearly don't have a good grasp of the situation in the real world. Like it or not, as long as the poor remain so obviously poor there will be resentment of the rich and a tendency for the leaders of the poor nations to be uncooperative with those of the rich countries when it comes to global political and economic agreements.

Paradoxically, it turns out that continued growth in the developed countries can help to stimulate growth of the poor, growth which, according to most economists, goes hand in hand with improved welfare and helps also to remove jealousy, bitterness and the scope for international friction. Unless the world as a whole continues to 'grow' in economic terms, the poor will stay poor unless and until the rich decide to give some of their riches away – or until the poor take them by force. It is arguable, of course, that the poor would be too weak to stand up to the rich in such a situation, and an argument is made by supporters of the concept of global 'triage' that the time has come to abandon the poor to their fate and concentrate our resources on saving civilisation as we know it in the rich countries. Such a 'solution' to the perceived problems might work; ultimately, there is a moral choice to be made in deciding whether or not the road towards global equality is the one to follow. But the evidence gathered by the SPRU team certainly does not suggest that resources are so limited that we *must* abandon the poor to their fate. My own choice – and for once I do speak here for the whole SPRU team –

is for the road to equality, and this subjective choice of a pleasant future world to live in narrows the area of debate still further.

The crucial aspects of any possible future world which might develop from the present situation are now seen in terms of only two variable quantities: the growth of world productivity as a whole, and the distribution of the fruits of that productivity between the rich and poor countries. Each variable may go in either of two directions compared with the present situation: more or less growth, more or less equality between nations. So we are left with only four kinds of possible future world to consider:
1. High growth, high equality (a big cake fairly shared)
2. High growth, low equality (a big cake hogged by the rich)
3. Low growth, high equality (a small cake, fairly shared)
4. Low growth, low equality (a small cake hogged by the rich)

Within each of these 'scenarios' there is still scope for investigation of detailed effects such as the preservation of a pleasant environment, the quality of life, and so on. But there is no plausible future world which cannot be fitted into one of these scenarios, as we can see from Figure 1 in which all the familiar, and some not so familiar, futurologists are put into the appropriate perspective.

The great debate about these four kinds of possible future world has formed the subject of the major SPRU forecasting effort described in *World Futures*. With the background of this intensive study behind us, we can concentrate in this book on the one scenario that emerges after study as not only possible but desirable, providing us with a pleasant future world to live in.

In 'commonsense' terms (commonsense is always easy with hindsight, once someone has done an indepth study and provided the detailed answers!) we can sketch out the implications of each scenario very briefly, when the

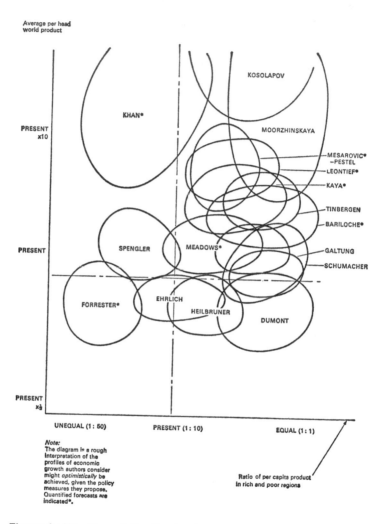

Figure 1 : Wealth and distribution in preferred futures

desirable option stands out like a sore thumb.

The first possibility, a big cake fairly shared, is such a welcome prospect that it has to be looked at seriously, and tested for the other factors affecting quality of life. Can we have a big cake, with something for everybody, and still not destroy planet Earth as a pleasant home? Or, in the phrase that has become familiar, can we achieve these aims within the limits to growth, whatever those limits really are? If so, then certainly the prospect is welcoming enough to justify a major effort to achieve it, even an effort which involves some change in the present world order.

The second prospect, of the rich continuing to get richer while the poor are ground in the mire, is so clearly a recipe for disaster in terms of global conflict that it can be ruled out of court. And the same can be said of our fourth scenario, which in selfish terms is an even better recipe for disaster since with only modest riches to hold on to the rich would be even less likely to be able to hold the poor in subjection by force, military or economic.

For the third prospect, however, with a belt tightening effort among the rich and fairer shares of the limited available resources, it is possible to find a wide body of support today, summarised by *The Limits to Growth* and *Small is Beautiful* evangelists. If we are now approaching the limits to growth, and must rest content (or at least make do) with a smallish global cake then clearly, in the interests of self-preservation and avoiding global conflict, we should share what little we have. But this implies an enormous strain on existing world systems, at least in the developed countries. Standing still, a no-growth situation, is one thing; but actually taking a cut in living standards, however worthwhile the reasons, is quite another. The hazardous implications of this kind of future world are spelled out in detail in *World Futures*. Here, it stands as 'obvious' that this is a situation of the last resort, only to be countenanced if all else fails. And since the main thrust of my argument, based on the SPRU work, is that all else has *not* yet failed, we are left to consider in depth the first choice scenario, the array of future worlds in which growth continues globally and the resulting riches are shared more equably.

High growth and more equality, then, are the twin themes of this book. But these are only crude overall indicators. More growth doesn't necessarily mean a better quality of life; concentration on closing the gaps between rich and poor *nations* doesn't necessarily imply an equable distribution of the fruits of production *within* any one nation; and so on. So it is by no means sufficient to sit back and say 'what the world needs is a bigger cake, shared more fairly'. We have to look at how this desirable end can be achieved in a future world pleasant to live in, and in order to do this the SPRU work has focused on three key issues: food, energy and material resources. Clearly, we have to eat, and there is every sign that the world's population will continue to grow, though not necessarily exponentially, during the fifty years or so covered by our look into the future; economic growth, necessary to create our 'best possible' world, surely requires increased use of energy; and if we are to have a bigger cake to share out then we most certainly need the basic ingredients to put into the cake, raw materials and, in particular, metals. Can these requirements be met in practice while maintaining – or better, improving – the quality of life? This is the central problem tackled in the middle part of this book; it is giving away no secrets to say that the answer we will arrive at is 'yes', since, of course, there would hardly be a book to write about the high growth/more equal* future worlds otherwise!

* The easy expression 'high growth/more equal' to describe a desirable future world is a distortion of the English language, bringing to mind the comment 'all men are equal, but some are more equal than others'. Strictly, I should refer to a 'high growth/less inegalitarian' world, but 'high growth/more equal' is a snappier way of saying something that will be repeated often throughout this book. For ease of reading, therefore, I shall stick with the snappy version, with the understanding that we interpret this as describing a shift which reduces existing inequalities, rather than making some people more equal than others. I apologise to any devotees of precise English whose sensibilities are offended in the process.

For much of the detailed analysis behind this work, the interested reader must inevitably be referred to the full background provided in *World Futures*. With that background, however, plus the information about food, energy and material resources discussed here, it becomes possible to paint a picture in the last part of this book of what it will be like to live in one or other of the possible high growth/more equal future worlds. In particular, perhaps to the surprise of some people, it turns out that this kind of scenario can be developed whatever your 'political' beliefs, and that 'conservatives', 'reformists' and 'radicals' can all envisage different paths towards this desirable end. To help readers who do wish to draw on the body of background information in *World Futures*, I have followed the same three-part structure as that book in presenting the story of high growth/more equal future worlds – though I stress, for the benefit of more trusting readers, that the present book stands alone if you are willing to accept my interpretation of that background material. And, logically enough, we set off on our journey into the future by looking at the broad features of the whole futures debate, defining the many forks in the road ahead from which the different possible future worlds may develop. Lest we should get too bogged down in detail along the way, however, I have one of my few direct quotes from *World Futures* to keep in mind throughout this book – perhaps the single most important conclusion to have emerged from the years of study put in by the SPRU team, a conclusion both encouraging and radically different from much of the gloom about the future which has made headlines in recent years:

'A desirable future is at least in the realms of the possible, if the right social, technological and political choices are made.'

Part One

Alternative Futures:
Where are we going?

Chapter One

Boom or Gloom? The Great Debate

Over the past few years the great debate about world futures – where are we going, and how are we going to get there – has thrown up two completely contrasting 'world views'. One view, labelled a 'doomsday' scenario and with proponents sometimes referred to as 'the prophets of doom' sees the world as we know it more or less inevitably falling apart and crashing about our ears, as unrestricted growth of population, runaway pollution and overexploitation of limited resources runs riot. This is certainly a pretty gloomy view: but it isn't, in fact, the worst thing that could happen, since there remains the almost unmentionable, in polite society, threat of all-out nuclear war. An even more dramatic possible end for civilisation as we know it. So I choose a more modest name for the eco-catastrophe doomsayers, the 'prophets of gloom'. The *real* doomsday stuff certainly is not going to be ignored in this book, and will be discussed fully in Chapter Two. But here I want to concentrate on the contrast between those prophets of gloom and the other extreme view presented by the opposed school of thought, those superoptimists best described as the 'prophets of boom'.

So far from seeing the structure of civilisation overheating and collapsing within a few years, the prophets of boom see improved technology solving all our problems, with growth of the world economy roaring ahead and ushering in the good times for just about everybody, perhaps as early as the next century. Boom or gloom? In futurology, it seems, you pays your money and you takes your choice! And, if the 'experts' can disagree so violently about the basic direction the world is going in, then non-experts can surely be excused if they decide no one really knows what is going on and ignore the whole debate.

But that would be unfortunate. As in so many areas of human thought and behaviour, the two extremes conceal a middle ground of likely developments, more likely than either boom or gloom in the extreme forms put forward so loudly by their prophets. Looking at the extremes helps us to understand what is going on in the real world, just as, by stressing aircraft components to destruction in the workshop, engineers help to ensure that real planes, subjected to less extreme stresses in 'real life', are safe to fly in. We don't have to accept the extreme viewpoints as gospel truth in order for them to be valuable in helping us to choose a best possible path into the future ahead. But, equally, unless you are prepared to devote a lifetime to the study of the futures 'game', there isn't much point in trying to examine in detail *every* theory thrown into the ring.

The debate goes back for centuries – millennia, even, if you include the ideas of the ancient Greeks – and today encompasses every shade of grey opinion from the black of gloom to the white of boom. In more or less modern times, the father of futures debate is usually regarded as Thomas Malthus, a nineteenth-century English economist who calculated that world population was growing faster than the rate at which food supply could be increased, and therefore forecast inevitable mass starvation. Such calculations are neat, simple to apply and dramatic in their implications. But they don't always end up being fulfilled in the real world, as another often quoted example shows.

Not long before the advent of the internal combustion engine as a power source for vehicles, nineteenth-century

London was beginning to run into severe problems caused by the necessity of feeding a large population of horses used to draw the many vehicles in the streets, and, no less important, removing the wastes produced by those horses. A simple Malthusian extrapolation made at the time could have shown that, as the number of horses increased, London would end up clogged into immobility by manure. The projection 'went wrong', of course, because horses were replaced by other forms of motive power. Today, one of the problems confronting further growth of society as we know it is the risk of increased pollution from those same internal combustion engines that saved us from a fate worse than death. But the optimists, the prophets of boom, would argue that technology can once again save us from the ultimate pollution hazards.

So this example highlights the difference between the two extreme 'world views'. From now on, I shall concentrate only on the recent revival of the futures debate, since about 1965, looking back not even to Malthus, let alone the Greeks. In addition, I shall try to avoid getting bogged down in the subtleties of shades of grey by looking just at the extremes – the gloomy forecasts of such writers as the authors of the notorious *The Limits of Growth*, and the wild optimism of the prophets of boom, headed by their archpriest Herman Kahn, riding in parallel, by a stroke of supreme irony, with the most widely publicised views of Soviet writers on the future.

Prophets of Boom

Historically, Kahn and his think tank at the Hudson Institute in the USA were the first of these futurologists to burst upon the scene, and he has the clear distinction of stirring debate, and looking at the prospects for the future, back in the early sixties. So let's pay him the appropriate tribute of looking first at the products of his school of thought.

Kahn's view of the good times just around the corner

appeared in 1967 in a book *The Year 2000* which emphasised the view that 'economic trends [i.e., growth] will proceed more or less smoothly through the next thirty years and beyond', although this book is one of the rare examples which does also look at the possibility of thermonuclear war. The world of the year 2000 portrayed in the absence of such disaster, however, is one in which all countries have become richer, and although some have got more rich more quickly than others, the basic world situation is like that today, only bigger (and by implication better). Not just the good times ahead, but 'western' good times – everyone building in the image of the United States of the mid-1960s. More hamburgers! More fast cars! More neon lights! More colour TV! and so on.

To be fair, the book also looks at other implications of the 'post-industrial' society developing in America, with the upper-middle classes living a life of ease like that of the landed gentry of yesteryear, increased life expectancy, health-improving devices, and other forms of 'progress'. The authors also see the continued existence of a very poor group within this sea of wealth, but a poor group that can be contained, fobbed off and prevented from bringing about sweeping changes to redistribute the wealth. The poor countries are seen as no threat to the rich, in physical terms, simply because they are too poor to fight effectively.

The book also pays lip service to the problems of technological change, including increased nuclear proliferation, but by and large it leaves the unsuspecting reader with a feeling not only that such a future is possible, but that it is pretty well inevitable. Several other futurologists find this sticks in the throat, and the view has been widely expressed that *The Year 2000* was really a deliberate attempt to influence public opinion and steer the world towards this possible future, rather than a serious entry in the futures game. One of the game-players, indeed, went so far as to call Kahn's vision of the future 'semi-lunatic'. But, lunatic or not, there is no doubt that the forecasting work of Kahn's team has been taken seriously and has influenced policy making by both government and big business in the US. The dangers are obvious – by presenting any one view as 'the future' we may make it inevitable, since we all begin to act in the way laid down by the blueprint. So perhaps it's just as well that the prophets of gloom came along to shake up the Kahnian picture.

It takes a lot to daunt a real prophet of boom, though, and by 1977, when *The Year 2000* was beginning to look a little outmoded (not least after the excitement in the early 1970s produced by the OPEC decision to increase oil prices dramatically – precisely the kind of effective action by less developed countries to hurt the rich that had been ruled out in the book) Kahn's team came up with a new contribution, in the wake of the US bicentennial, called *The Next 200 Years*.

It would take a brave opponent to describe this latest work as even 'semi-lunatic', and the detailed comments made in the new book generally strike even the critical reader as very sane. Somehow, though, Kahn has retained his image as the archpriest of boom while actually modifying his views to bring them much closer to the mainstream of plausible future worlds. This is particularly interesting, since as we shall see shortly the prophets of gloom have made a similar shift, away from gloomy extremism and closer to the middle ground, in the past couple of years. Meanwhile, however, their disciples remain at loggerheads with the disciples of boom, both groups having failed to notice how close their leaders are getting to speaking the same language!

Only an extreme gloommonger, now, could take great exception to such comments as:

In our view, the application of a modicum of intelligence and good management in dealing with current problems can enable economic growth to continue for a considerable period of time, to the benefit, rather than the detriment, of mankind. We argue that without such growth the disparities among nations so regretted today would probably never be overcome, that 'no growth' would consign the poor to indefinite poverty and increase the present tensions between 'haves and 'have-nots'. Nevertheless, we do not expect economic growth to continue indefinitely; instead, its recent exponential rate will probably slow gradually to a low or zero rate. Our differences with those who advocate limits to growth deal less with the likelihood of this change than with the reasons for it.

The 'reasons for it' put forward in *The Next 200 Years* revolve around the suggestion that we are now living through a great transition in the history of mankind, following the industrial revolution. Not exponential growth, but growth following a flattened 'S' curve is the picture favoured by the Hudson Institute (see Figure 2), and the way out of the dilemma is simply to wait and let nature take its course. The Industrial Revolution is put in perspective as a watershed in history comparable to the development of agriculture some 10,000 years ago, a change so dramatic that we are still adjusting and cannot expect society to settle down into a new, stable pattern for another 200 years or so – which means that, by luck, we just happen to be living through the most exciting period of change in world history (whether the luck is good or bad depends on your own philosophy and personal world view!). Exponential growth says Kahn and his colleagues 'does in fact appear to be stopping now, and not for reasons associated with desperate physical limitations to growth'.

The new catchphrase is 'the demographic transition' – the transition which has occurred in every country as wealth has increased, life expectancy has gone up and quality of life has improved, which brings a corresponding decline in the birth rate. Not only has the transition from high growth rates to low occurred in every developed country, but the transition is getting faster as more of the world moves through the transition. In western Europe and the USA the change took 150 years; in the Soviet Union, forty years, and for Japan, a recent case, only

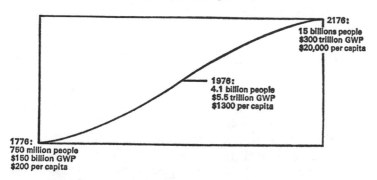

A Bicentennial and/or Realistic Perspective on
Prospects for Mankind
(in fixed 1975 dollars)

2176:
15 billions people
$300 trillion GWP
$20,000 per capita

1976:
4.1 billion people
$5.5 trillion GWP
$1300 per capita

1776:
750 million people
$150 billion GWP
$200 per capita

Figure 2: The great transition (from *The Next 200 Years*)

twenty-five years, from 1935 to 1960. Economic growth and the transition from underdeveloped to developed status is also seen as speeding up in *The Next 200 Years* – it's as if the process gets easier as the trail blazed by western Europe and the USA is beaten into a broad path by the following nations and world regions.

All this is heady stuff, and indeed more soundly based, though still optimistic, than *The Year 2000*. Now, the year 2000 is being offered as not quite the good times for all, although moving that way; equally, the nature of development is no longer seen as exclusively 'western' good times. For the range 50–100 years ahead that is the focus of attention of the present book, one last quote from *The Next 200 Years* is particularly apposite – and, as we shall see later, not all that contentious:

By the year 2000 perhaps a quarter of mankind will live in emerging post-industrial societies and more than two thirds will have passed the level of $1000 per capita [product per annum]. By the end of the twenty-first century almost all societies should have a GNP per capita greater than $2000 and be entering some form of post-industrial culture. *The task is not to see that these*

31

societies proceed along the same path as Europe, North America and Japan, but rather that each should find its own way. [my italics]

The significance of the figures quoted, modest though they may seem to most western eyes, is well known to economists. Moving from $100 to $1000 per capita* is sufficient to take a country from abject poverty to present day 'lower middle-class' standards, and in real terms this is far more important than growth from $1000 to $10,000 per capita, although both represent a tenfold increase. In Kahn's future world, the choice is 'not between poverty and desperate poverty, as some pessimists have suggested, but rather between failure and success, in which "failure" means an annual per capita product of $500 to $2000 for the poorer countries and "moderate success" means a range of $3000 to $10,000'.

However else his views may have moderated over the ten years between publication of the two books discussed here, Kahn's vision of growth remains essentially 'capitalist' in the broadest sense. So it is ironic, and worthy of at least passing mention, that the other location of a major group of prophets of boom is in the Soviet Union – where the reasons for sustained growth as a solution to world problems are seen in quite different terms!

To Soviet eyes, the problems envisaged by western gloom-mongers following the example of Malthus, the neo-Malthusian 'collapse' scenarios, are seen quite simply as inherent weaknesses of the western capitalist-imperialist system. The poor, developing nations are now seen to be poor, with inadequate food, basically because they are oppressed and exploited by the rich. The answer is equally clear – socialist revolution to remove inequalities, provide fair shares for all, and lead to a future world of technologically based growth – the 'good times', once again, but now definitely Leninist-Marxist good times!

Whatever the route to the future, however, the visions

---

* At fixed prices, that is, taking account of inflation. Throughout this book, unless otherwise specified, the chosen fixed price 'base' is 1970.

portrayed by Kahn and some Soviet writers are more remarkable for their similarities than their differences. Modrzhinskaya, for example, writes of 'a considerable rise in the material and cultural standards of the whole people' of the Soviet Union by about the year 2000, with in some cities 'ultra-high [living] blocks of 200–300 storeys', while Kosolapov suggests that by 2005 AD tourist trips to the Moon will be available.

No, the place to look for contrasts in the futures game is not between East and West, but *within* the western world, where Kahn's road to the future is so strikingly different to that of the neo-Malthusians.

Prophets of Gloom

The great thing about *The Limits to Growth* is that its publication early in the 1970s sparked off a great wave of debate about the future. But one unfortunate side effect of this public attention is that to most of the people who are aware of 'the futures debate' the debate about *The Limits to Growth* and the neo-Malthusian gloommongers *is* the 'futures debate'. But just as Kahn was thinking about the future fairly publically, if without gaining quite so much attention as the Club of Rome, long before the arrival of *The Limits of Growth* on the scene, so there were forerunners of gloom leading up to what we might call 'the limits debate'.

If anyone can be credited with parentage of the idea of eco-catastrophe, in its 1970s form, it is surely Anne and Paul Ehrlich, who argued in several books published in 1970 and 1971 that the world was even then *already* 'over the top' in terms of population and destruction of resources. To them, 'Spaceship Earth' is already seriously crippled by runaway growth and unthinking pollution; famine, plague and nuclear war are seen as the only likely future worlds, and the only hope, so far from suggesting more growth, is to start taking things to bits before the system falls apart in our hands.

This is, essentially, what I referred to in the introduction as the 'low growth/more equal' scenario. The rich, according to the Ehrlichs, must as a matter of urgency give away their worldly goods to the poor in the hope of ultimate redemption. 'De-development' is the watchword – not just for the west, but for the Soviet Union, which is seen as an equally villainous despoiler of the environment. The basis for such a dramatic turnaround in the way society acts could only involve a change in moral attitudes, the kind of change usually associated with religious conversion, and it is no coincidence that the Ehrlichs preach their case in such terms. Their message is indeed a familiar one 'repent, for the end of the world is at hand', and as such it is telling us more about human nature, perhaps, than about the true prospects facing the world today.

Sitting rather uneasily alongside this highly moral contribution to the futures debate comes the claim from the Ehrlichs that things have reached such a desperate pass that for some very poor nations there is already no hope. Their solution? Triage – a nice, clean jargon term for the rather dirty prospect of writing off whole nations who are beyond 'realistic help' and leaving them to suffer while we save the rest of the world from disaster. This surely marks the ultimate extreme from the super-optimism of Kahn, a counsel of despair so bleak that alongside it even *The Limits to Growth* looks not just sane but almost cosy and respectable.

If the Ehrlichs were right, we really might as well dig a pit and bury ourselves in it now; other gloommongers, however, through the process of the great futures debate, have shifted, like Kahn, towards the middle ground where a path towards a reasonably pleasant future world can already be discerned.

Before this moderating process, however, the prophets of gloom had quite a field day. And what made people sit up and take notice was the arrival of computer modelling on the scene of the futures game, as if running numbers through a computer made them infallible as a guide to the future (a reaction which is equally interesting as a pointer to human nature as the Ehrlichs' contribution – but one

34

which ought to be taken with a pinch of salt by anyone who has had a run-in with a computerised bank statement or airline booking system!). The neatest description of this new contribution to the futures game is 'Malthus with a computer' – a tag which makes the point that what comes *out* of a computer model depends on what you put *in* – the background theory and, if you like, 'philosophy' of the computer programmers.

Put Malthusian ideas into a computer and, wonder of wonders, out come Malthusian ideas magnified a hundredfold, and printed on impressive-looking computer paper. Newspapers and many non-specialists gobbled up the authoritative output from the computers without pausing to question what had gone in – rather as if you received a computerised bank statement showing your account to be £1,000,000 overdrawn, and accepted this as gospel without asking how the numbers had been arrived at. The importance of the work by Jay Forrester, who first developed the computer programs, and Dennis Meadows's team, who used them to great effect, lies not in their output alone but in the way this output depends on what is fed in to the computer. If we look at how things change if the *assumptions* we are putting in are changed, then we might really gain some useful understanding of how things work in the real world.

Take one example. A key feature of the 'limits' work is that it used global averages (for growth, population, pollution and so on). Yet the biggest problem facing the world, in the view of the SPRU team and many other experts, is that of inequality. How can you begin to look at the effects of inequality – even the desperate effects implied by the work of the Ehrlichs – if your 'model' of the world averages everything out? Isn't a 'global average' of everything more or less what the Ehrlichs were advocating, anyway?

And the numbers that come out of the Meadows's team's calculations also look curious in many other ways – not least since, while portraying a world of short-lived rapid growth followed by dramatic collapse, the growth figures for the brief boom period are a great deal higher than those used by Herman Kahn, the archpriest of boom himself! The proposed solution to the problem is equally curious – an end to growth, but with no suggestion at all about how to resolve the problems of the poor nations in a 'no-growth' world. Not even 'low growth/more equal', but apparently 'low growth/status quo' is the outcome of the Forrester-Meadows future world – and just how long would the poor stand for that before deciding to grab what they could?

Robert Heilbroner has drawn the logical conclusion from accepting the 'limits' work at face value: 'only two outcomes are imaginable - "anarchy" or "totalitarianism"'. It says something of the despair induced by taking these computer models as gospel that Heilbroner then accepts totalitarianism as the lesser of the two possible evils, and that there are many who would agree with him. So perhaps, before moving into the middle ground proper, we should take a slightly closer look at those 'infallible' computer forecasts.

The Controversial Computer

Although the computer's initial impact came chiefly through its apparently impartial presentation of 'the facts', the true role of computer models in the futures debate is as an integral part of the 'political' aspect of the game. With a good computer model as camouflage, the skilled programmer can hide the political bias going in to the game – but equally, another expert at the game can strip away the camouflage to lay the bias bare. The SPRU bias, which I share and which this book is all about, is that of achieving a more equitable distribution of the 'good times', both between countries and within them. To achieve this, it seems that there must be some kind of growth and technological progress. The gloommongers, with or without their computer models, are as much neo-Luddites as neo-Malthusians, arguing that technology

36

is bad, or at best is reaching the limits of the possible, and that growth must stop first *before*, if ever, we tackle the problems produced by uneven distribution in the world. What the computer provides is a way to try out different kinds of bias and see what 'predictions' emerge. The modelling is a game in which we can change the rules to some extent, in order to find out which set of rules are going to be most useful in the real world. And that is where the computer is invaluable, since it speeds the game up and makes it possible to try out many different ideas in a reasonable space of time.

The best way to show this is to try changing some of the 'rules' and playing a slightly different 'game' with the Forrester-Meadows models. This is just what the SPRU team did in the mid-1970s (see *Thinking About the Future*), with results that are entertaining enough to be well worth describing briefly here. The starting point is the 'standard run' of a computer model called 'World 2', which produces the curves shown in Figure 3, the dramatic boom and collapse scenario of the 'limits' debate.

Now, you have to start your model 'running' from some time when things like birth rate and death rate are pretty well known – that's one of the rules of the game – and you have to make various assumptions about what is happening to the world economically, in terms of technological change, and as affected by pollution – those are the other rules of the game. Without bothering about just which rules the 'limits' team put in, 1900 seems a sensible enough place to start the run, appealing to the human preference for nice round numbers. But although this 'run' gives curves which match up with reality fairly well from about 1910 to 1970, the death rate figures are a bit 'off' in the years just after 1900. One of the delights of computer models of this kind is that they can be run backwards as well as forwards, and this is a good test of their stability – a really good model should run backwards and forwards between chosen dates and always produce the same curves. So what happens if we run the World 2 model, the key

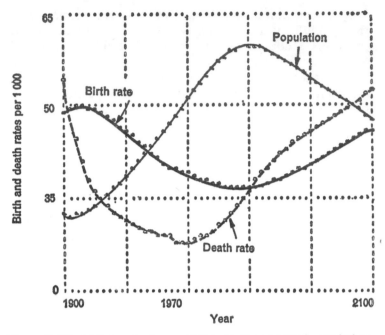

Figure 3: World 2 standard run: births, deaths and total population changes

model which gave impetus to the publicity for *The Limits to Growth*, backwards from 1900?

The result (Figure 4) is curious, to say the least. According to the model, the world we now live in is a product of a dramatic collapse in the late nineteenth century, when world population slumped from some almost unimaginable peak, *greater than world population in 1970*, prior to 1880! The great 'catastrophe' of 1880 is so dramatic that the computer model can't be run back 'through' it – as far as World 2 is concerned, with that particular set of rules, that was the beginning of the world.

This kind of behaviour is just the thing that points up a need to change the rules of the game in order to get the machine to play properly. So let's change them and see

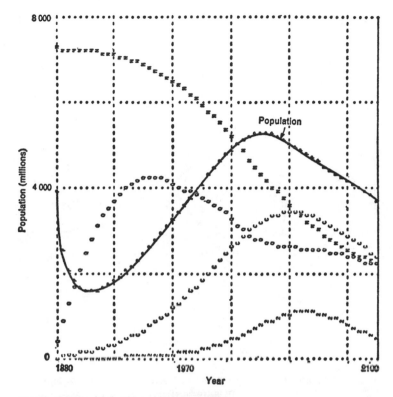

Figure 4 : World 2. History from 1880

what happens. Taking only one of Forrester's assumptions, about the link between material standards of living and death rates, the 1880 catastrophe can be avoided. All that the SPRU team did was to alter Forrester's *guess* that changes in average world material standards can cause death rates to vary by a factor of six (×6) to their own *guess* that the true variation is by a factor of three (×3) to produce the curves of Figure 5, where the model runs smoothly back to the early nineteenth century.

The new curves clearly bear a much stronger resemblance to known history – and now, although population declines in the twenty-first century the *reason* for the

39

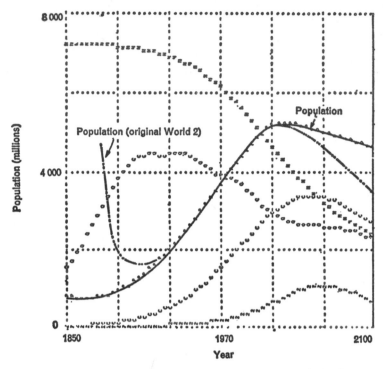

Figure 5: World 2. History from 1850 with a changed death rate assumption

decline is a falling birth rate not, as in the earlier version of the game, an increasing death rate! Many of the other guesses put in to the model by Forrester and others can be adjusted with equally dramatic effects. Quite reasonable estimates of agricultural productivity, but different from Forrester's guesses, change the picture by changing availability of food, and so on. What we see is that with neo-Malthusian, Forrester-Meadows gloomy guesses about how the world works we get gloomy forecasts; with more optimistic guesses we get more optimistic forecasts. Rather than saying 'the end is inevitable', the model is actually giving us a clear indication that we have a choice

40

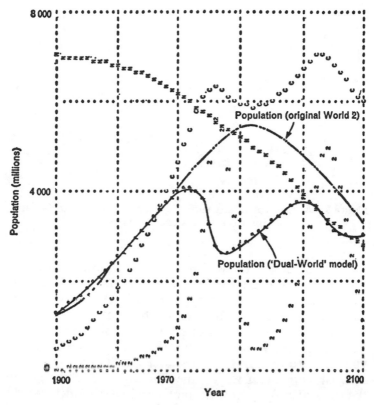

Figure 6: World 2. Dual-world version, with pollution restricted to region of emission

of future worlds and that the outcome depends on our actions.

The naivety of World 2, however, is shown when we split the world into developed and underdeveloped regions (the best we can do with this model to simulate the inequalities of the real world) and run the model with the assumption that pollution produced by industrial activity stays more or less where it is created. Now (Figure 6), the developed world has its own catastrophic collapse (starting in about 1972, but let's not pick over details) while the developing

41

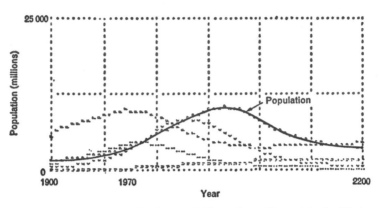

Figure 7: World 3 standard runs with scales adjusted to facilitate comparison with succeeding graphs

world proceeds blithely along the same path until it too suffers an eco-catastrophe. Of course, *if* the developed world fell apart in this way, the world system would be changed so much that the rest of the world would follow a different path into the future. But computers are stupid – *we* can see this is obvious, but the machine won't unless you tell it to change the rules after the first collapse.

What happens if we go the whole hog and change *all* the rules? Looking now at another, similar model, World 3, the SPRU team compared the standard forecast of gloom (Figure 7) with a variation in which the assumptions about resources, pollution, agricultural and capital distribution are all changed towards the optimistic end of the range of possibilities that has been seriously discussed in the futures debate. With optimistic, but not insane, assumptions the result is dramatic – we are back to the Kahn vision of the future! This, of course, is just what we should expect – if 'Malthus in, Malthus out' is a truism, then it is equally certain that we can apply the dictum 'Kahn in, Kahn out'. The computer is an idiot savant – it does what you tell it, very quickly, but *only* what you tell it.

42

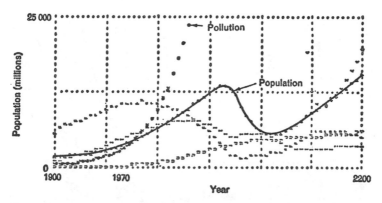

Figure 7a : World 3 with changed resource costs

The shift in position of Meadows's team, mentioned before, seems to highlight their realization of this. In *The Limits to Growth* itself, readers were told that all of the conclusions were borne out by detailed work which would be made available later as a Technical Report. Unambiguous statements such as 'the limits to growth on this planet will be reached in the next hundred years' and 'the basic behaviour mode of the world system [is the same] even if we assume any number of technological changes in the system' provided the cornerstone of their disciples' faith in the furore which followed publication of *The Limits of Growth*. But what happened when the Technical Report became available (without so much publicity and attention as the 'popular' book)? Now, the comments are more in line with the game-playing results described here, and a key comment is that

As the runs presented [in the Technical Report] have shown, it is possible to pick a set of parameters which allow material, capital and population growth to continue through the year 2100.

In spite of this, the disciples have, if anything, hardened their entrenched positions and are still to be found

43

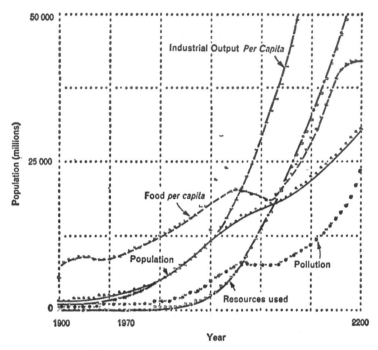

Figure 8: World 3 with changed resource, pollution, agricultural
and capital distribution assumptions

quoting as gospel 'the limits to growth . . . will be reached
in the next hundred years', although Meadows's more
moderate comments were available, to those in the know,
in June of 1972!

This is, perhaps, a good place to draw attention to the
fact that my own subjective interpretation of the state of
play in the futures game is derived not from some un-
published or difficult to find technical report, which might
one day appear to vindicate my statements (or not, as the
case may be) but on the already published work of the
Science Policy Research Unit, books which are available
to anyone who wishes to check the background to my
story. This is, after all, the more usual way of doing

44

things – technical material first, subjective interpretation later. Rather than saying 'these are the opinions which my facts are going to support' I prefer the old-fashioned approach – 'there are the facts on which my opinions are based'.

But all of the futures debate from 1965 to about 1975 was merely setting the scene for the main work. The extremists have established their positions for debate, then shifted position slightly, like boxers sparring before the square-up for a slogging match. World 2 and World 3 were recognised by their users (though not, alas, by their disciples) as the first crude attempts to find out where the world is going. 'Second generation' computer models are now beginning to take account of differences between regions, instead of using global averages, and whereas the original Forrester model had forty equations in it, and the Meadows version 200, today the modellers talk in terms of 100,000 equations to describe the world system.

Of course, even with 100,000 equations (or 1,000,000), if you still put Malthusian ideas in you'll still get Malthusian gloom out, but you can begin to play different games, and learn different things about the world system, if you want to.

As we have seen, not everyone relies on an electronic computer, some forecasters carry their models in their heads. The 'mind modellers' also are moving into a second generation version of their work, as the distinctions between *The Year 2000* and *The Next 200 Years* show. And it is within this ferment of second generation ideas and work that the SPRU contribution, with the prospects for a high growth/more equal world, really belongs.

The Second Generation and the SPRU Approach
Far more important than the number of equations used in the models is the emergence in the second generation work of studies which take account of the different regions of the globe, and even the distribution *within* regions,

instead of using simple global averages. The Latin American Bariloche team is one which has adopted this approach, and which sees clearly the implication that the less developed parts of the world today will *not* develop in exactly the same fashion as the already developed nations, but through socio-political factors rather than through the limits of natural resources. In this future world, collapse is seen as a real possibility, not through Malthusian over-growth, but because of the instability produced by a widening gap between rich and poor – the 'haves' and 'have-nots'. The scenario fits within the overall picture of 'more growth/less equal', perhaps the most explosively unstable future world of all. As for the Club of Rome, sponsors of *The Limits to Growth*, their most significant second generation contribution to date has been the work published as *Reshaping the International Order* (RIO), produced by a team headed by Jan Tinbergen. The moral tone of their viewpoint shines through clearly in the RIO work, which calls for a world in which 'a life of dignity and well-being becomes an inalienable right of all', and which, no less interestingly from our present point of view, comments that 'fears expressed in recent years concerning the exhaustion of natural resources may well be exaggerated'. The RIO study (which, by the way, does *not* use a computer model!) sees the Club of Rome moving squarely into line with the view that although there are grave problems facing the world they can be tackled effectively and solved; that we have a choice of future worlds depending on our actions in the face of these problems; and which stresses the moral virtues of moving towards a more equitable society.

Where does the SPRU work fit in to all this? Like the RIO study, and the work by Kahn's group, their approach depends not on detailed number-crunching in a computer but by building up scenarios around a framework of explicitly expressed assumptions. The four basic scenarios have already been met in the introduction, and we have seen how the different schools of thought in the futures

debate can be accommodated within this simple framework. The computer models present 'numerical' views of
the future – but the exact numbers are much less important
than the broad patterns, which they tend to disguise, like
that famous wood that couldn't be seen for the trees.

The models also disguise the nature of the underlying
assumptions put into them, as we have seen, and it is an
important part of the SPRU approach that assumptions
and the implications of those assumptions can be clearly
seen and related. Since the basic assumptions made are of
key importance in determining ultimate trends, the use of
computer models is hardly necessary – we can tell anyway
that if we assume world resources are very limited then
any increase in consumption will bring disaster. What is at
issue is not the link between assumed cause and ultimate
effect, but the basic assumptions themselves. Are the
world's resources really so limited? Is it really impossible to
feed a population twice that of the world today? Can
enough energy be provided to power an industrialised
twenty-first-century world? With suitable answers to those
questions – answers established from looking at the world
as it is now and as it has changed in recent decades – we
can be sure that the world can grow and produce the
means to reduce inequalities. With the 'wrong' answers,
we wouldn't need a computer to paint a bleak picture of
the future.

But the existence of a possible desirable future world
does not of itself mean that we will get there. Choosing the
right path out of what is a pretty messy situation today is
the most important task of all. The worst thing about the
first generation futures work was not so much the dichot-
omy between 'growth is good' and 'small is beautiful' as
the lack of any guidance on choosing a path into the
future. Just to say 'keep on growing' or 'stop growth now'
isn't much use to the policy makers. Much better than
trying to forecast where the world is going is to look at
how particular issues that exist now, and may arise later,
should be tackled in order to steer society towards the best

47

possible future world. Even if we can't yet tell how each policy choice will act to shape the detailed long-term future (that must wait for the third, or later, generation approach) we can show why some factors are more important than others and should be looked at in the decision-making process. At least then if we make the wrong decisions and are led up a blind alley into a gloomy future world we can have the masochistic delight of knowing it's all our own fault!

Food, energy and raw materials are without doubt the three key issues at the heart of the whole futures debate, and are the issues which must be examined most carefully in this policy-oriented approach. Our own 'morality', our subjective view that a more equal world – or, at least, a world with less inequality – is a desirable future, provides the driving force behind the examination of these key issues; but even if your own inclination is to favour a rigidly structured system in which the poor are oppressed by the rich you would still have to look at the same issues in order to decide how feasible such a world might be.

So it is the discussion of those three key issues that lies at the heart of the present book. With our place in the continuing futures debate established, it is almost time to get to work on the heart of the matter; but first, we have to face up to the grim possibility of a future in which all rational bets are off – the prospect of all-out war.

Chapter Two

Prospects of Doom

The possibility of all-out war – the real prospect of doom hanging over society – is something that is largely ignored in futures studies and, indeed, in polite conversation. This seems to be part of the pattern of human nature, to shy away from the most horrible problems in the hope that if we ignore them they might go away. The attitude is typified by the problems experienced by author Bruce Sibley in trying to find a publisher in Britain for a book which stresses the real prospects of nuclear war (*Surviving Doomsday*, Shaw & Sons, London, 1977), problems spelled out in an article in the *Guardian* on 21 November 1977. When Sibley showed his manuscript to a literary agent, he was told that:

> There are certain subjects which the reading public tends to shy away from – such as the fear of cancer or blindness. I am afraid one has to add an atomic holocaust to that category. This is purely a subjective opinion, but I seriously doubt whether readers would want to be both depressed and scared to that degree.

If such an attitude really is widespread – not just among general readers of books but among futurologists themselves, it is doubly unfortunate. For, in the first place, there is, after all, no doubt that the prospect of war is a real one, and that it could produce such a dramatic change in the pattern of human activity – a discontinuity –

that all our forecasts of the future would cease to have any relevance at all to the real world. Secondly, and perhaps of even greater importance, by lumping all prospects of war together as nuclear Armageddon, and deciding that the prospect is too horrific to contemplate, the unpleasant potential of present day, and likely future, 'conventional' arms and warfare is overlooked. Even if nuclear war was such a ghastly prospect that it didn't bear thinking about, we should think very seriously about what might be done under the reassuring label of 'conventional' warfare.

And there is, still, another face to the problem which appears more welcoming but which could mark an equally dramatic discontinuity in the development of our future world: disarmament. Preparations for war have been a part of human society for so long that what we now like to call 'defence' activities have become an integral feature of the structure of society. If we could move towards a future of disarmament, without this basic feature of society as we know it present, the change – the discontinuity – could be as dramatic as the effects of war themselves – although, we would hope, rather more pleasant. Ultimately, perhaps, a major discontinuity in the development of human society will be inevitable. If 'defence' spending continues to be a major use of resources, then sooner or later, by accident or design, war must come; if spending on armaments and military preparedness is stopped, then the whole pattern of society will change. Are these long range indications likely to be important on the timescale we are interested in, roughly up to a hundred years from now? The discontinuity implicit in a disarmed world hardly seems feasible on that sort of timescale, but while we may hope that all-out war can also be avoided, how long can preparation for war continue without war? Even without all-out war the presence of a well-fed war machine is itself certain to play a large part in deciding how we progress along the road into the future.

Feeding the War Machine

The demand of the war machine for 'food' in the form of a large and growing proportion of the available resources of the world (the 'world product') is reminiscent of nothing so much as a cuckoo, pushing its foster siblings out of the nest and exhausting its foster parents with its demands. Just before the First World War, with rearmament proceeding apace, about 3½ per cent of the total world output was going to military uses; by 1968, the proportion had more than doubled, rising above 7 per cent. At this rate of growth, by the year 2000 the consumption of the war machine will exceed the *total* world output of 1968! With military expenditure beginning to rise again from a plateau reached in the early 1970s, there seems every prospect of this remarkable forecast being realised; already, in the late 1970s, the spending on 'defence' around the world amounts to the equivalent of:

The *combined* production (Gross National Product) of all sixty-five countries that make up Latin America and Africa: or the *total* worldwide expenditure by governments on education; or *twice* the total worldwide expenditure by governments on health: or about *fifteen times* the value of all government aid to the underdeveloped countries combined.

Hardly surprisingly, the rich nations dominate this picture of a militaristic world. In 1975, expenditure by the US and USSR made up nearly two thirds of the total world military 'budget', with the allied members of the NATO and Warsaw Treaty groups contributing a further fifth. But, in contrast, the rate at which military expenditure is growing has been greatest lately in the poorer countries – a highly significant burden on nations already struggling to 'catch up' with the rich.

There is a sense, though, in which it can be argued that all this military activity is a good thing, because it encourages research into science and technology, and, in

51

addition, by providing work and spending money, provides a stimulus to the economy and a means to distribute available wealth. To most people, the 'commonsense' view would surely be that there is something wrong with an economy which can only be stimulated effectively by preparations to kill people; unfortunately, economists who argue such points generally avoid such emotive expressions as 'killing people' and seem to have forgotten what the war business is all about. Nevertheless, we need to have a look at some of these arguments, and how the war machine fits in to the overall economy of a modern country, before we can begin to think about its influence on future worlds.

There's no denying that the expenditure by governments on research and development (R&D) for military uses has been a very important contribution to the growth of science and technology in the twentieth century. Although it is widely appreciated that war itself – in particular the two World Wars – stimulated such development, with obvious examples being in the dramatic progress made with aircraft, communications, and, of course, weapons, many people still do not appreciate how much 'the military' has contributed to scientific and technological progress in the rich industrialised countries of the world since the end of the Second World War. For more than thirty years now, as a result of the world pattern that emerged after that war, the rich countries have felt a need to keep on developing weapons as rapidly as if there was a 'shooting war' still going on, for fear that without such 'defence' they would invite a takeover from outside.

One result of this activity has been that whole weapons systems – sophisticated methods of delivering nuclear warheads on target, and defence systems to counter such capabilities – have been developed, made obsolete and scrapped without ever being used. This is wasteful enough in itself, unless it can be proved that only the existence of those unused weapons really did prevent a disastrous war. But the effects of this frantic activity in military fields has been felt far wider, since it has set a pattern for the way science and technology have been applied and developed outside the military sphere.

Entirely non-military resources have been devoted to exotic, technologically complex and 'sophisticated' projects such as the development of the Concorde supersonic aircraft or successive generations of nuclear reactors. If you *want* nuclear power, does this approach – a striking echo of the military approach of making systems redundant as soon as they are workable – really indicate the best path to follow, or should we stick with one proven design? Do we need nuclear power and Concorde at all? If the same resources had been devoted to cheaper forms of transport or alternative forms of energy over the same period of time then we might not be in the mess we seem to be in now.

There's also another effect of expenditure on military R&D, charmingly termed 'the acorn effect', which helps to encourage the increasing levels of military expenditure. As things stand today, our scientific understanding and technological capabilities are such that with a little more effort we can produce more, different, bigger, better and more expensive weapons systems. A relatively small R&D budget is likely to be enough for work which proves a new development to be a practical proposition – and once that is so, in many cases the temptation to spend a great deal of money putting the new development into general practice in the field becomes overwhelming. Great trees of military expenditure from little acorns of R&D may grow.

So the growing military share of available resources eats up capital which might be applied elsewhere, as well as diverting scientific progress on to paths which might otherwise be avoided. By 1971 in the UK, arms expenditure accounted for a fifth of the mechanical engineering industry, a third of telecommunications and electrical equipment, half of shipbuilding and three-quarters of aerospace. All very well if we had a big enough cake to cope with the needs of society from the bits left over from military uses, but this just isn't the situation in the real world. It has even been suggested that the huge impact of military spending on these kinds of capital-goods industries in Britain and the US has, at least partly, caused the dramatic weakening of the relative trade performances of those two countries compared with those of Japan and West Germany since the early 1950s.

In the US, another effect of the growth of the war machine is shown by its increasing tendency to store up material against a rainy day – a kind of Parkinson's Law along the lines of 'the more you get, the more you want'. At present, more than a quarter of US military expenditure goes into procurement, and this proportion has been growing, so that procurement is taking an even bigger share of an ever bigger cake. This has its own inbuilt acorn effect – the more material you tuck away against that rainy day, the more warehouses you need to keep it in, the more men you need to look after it, and the more computers you need to keep track of exactly what you have got, and where. The more the military gobble up, the less there is for anyone else. The more procurement there is, the more military expenditure; and the more military expenditure there is, the less there is for civil investment to stimulate economic growth.

The logic is simple. In the case of the rich countries, at least, a highly militarised future world must be a low-growth world. The counter-argument, that military spending stimulates demand and boosts the economy, really doesn't stand up to close inspection, and it would be a pretty weird system – thoroughly deserving of a discontinuity! – which could only gear itself up for action by wasting a large fraction of its resources.

For the poor countries, though, these arguments do not apply in quite the same way. The weapons and equipment gobbled up by their military machines are largely imported, and the overall balance of the military budget is on man-power, not technology. However, one line of thought,

54

developed by (among others) Mary Kaldor of the SPRU team, relates high military expenditure in poor countries with the perpetuation, and even reinforcement, of unequal distribution of wealth; something rather like the stereotyped view of a comic-opera 'banana republic'. This is particularly important in trying to understand which path the world may take in the immediate future, since by combining this evidence with the clear evidence that extreme militarisation of the rich countries would ensure a low growth world we can see that militarisation leads directly to both low growth and unequal distribution. The debate about military spending and its relation to economic growth in the less developed countries continues, with the extreme positions being held by those who argue that such spending causes growth and their opponents who hold that it doesn't. The most plausible view seems to be that while such spending certainly may encourage growth, it also entails inequality and repression, thus preventing 'development' in the full sense.

We have already pondered in the Introduction on the likelihood that a low growth/unequal world would be one of tension and military friction; now, we see the feedback loop completed. Once such a situation became firmly established, it might be very difficult indeed to break out of the vicious cycle of military expenditure, lack of growth, and inequality, with, perhaps, localised wars helping to keep the cycle going while dragging down the standards from, perhaps, low growth to no growth, or even negative growth.

Three views of a militaristic future

How does the increasing war preparedness of the world, then, play a part in defining our range of plausible future worlds? Before we can really come to grips with the possible range of alternative futures we have to face the fact that there are almost as many different political perspectives, or 'worldviews', in the futures game as there are players. Everyone has their own idea of what is socially just and of how things 'ought' to be done in an 'ideal world'. In terms of the labels used by economists, the views of a particular futures forecasting team may be Marxist, Keynesian or Neo-classical; in the slightly more emotive terms of everyday usage, different forecasters may be dubbed 'capitalist', 'revisionist', 'socialist' or any of a variety of rude names. And, clearly, there is no way in which we can begin to analyse all of these different standpoints in the futures game (let alone the hairsplitting differences between, say, different *kinds* of Marxist) without getting swamped in the process and losing our grip on the real problem of where the world is going.

But all is not lost. For simplicity, the SPRU team has looked at three broad types of worldview, which they label 'conservative', 'reformist' and 'radical'. Together with the four kinds of future world implicit in the 'choice' of high or low growth, more or less equality, this gave them twelve full scenarios to play with, but, fortunately, we can leave even this detailed background aside and look, in due course, only at the different ways in which the best possible future, high growth/more equal, can unfold in each of the three worldviews. For – and this, perhaps, is as important as anything to have emerged from the SPRU study – it does turn out that such a desirable future can be achieved within any of these 'political' frameworks. The implication is that you would need a very odd, or extreme, politically flavoured future world for the desirable high growth/more equal option to be impossible – which, we must remember, is *not* the same thing as saying that the desirable path will *inevitably* be followed in the real world.

Although these three flavours are not simply 'left', 'middle' and 'right' in the conventional political spectrum, and the subtleties of their standpoints are explained at some length in *World Futures*, in simple terms it helps to think of them as, broadly speaking, traditionally capitalist *laissez-faire* economy (conservative), Keynesian intervention in the economy (reformist), and full socialist state control of the economy (radical).

To all the economists about to throw this book away in disgust, and my colleagues who worked so hard to get away from such simplistic labels, I apologise! But unless you are a trained economist it makes much more sense to work with everyday terms which we all have some kind of 'feel' for, and, after all, this is not supposed to be an economics text book.

Already, even with this simplistic interpretation of the three worldviews, the broad patterns of the paths towards different types of future can be uncovered. The important thing, however, remains not the political flavour of your preferred future world, but the fact that the alternatives (and especially the high growth/more equal alternative) can indeed be found by each route, and come in each political flavour.

If this seems surprising, remember that we have already come across the irony that the booming future envisaged by Herman Kahn's team (thoroughgoing conservatives in terms of our three flavours) is most closely paralleled by the rapid growth portrayed in the Soviet scenarios (definitely radical). While, in the 'middle ground' if you like, it's easy to find widely disparate views all sheltering under the reformist umbrella, ranging from Malthusian gloommongers across to some of the disciples of boom. Confusing, isn't it? But after all, this labelling of world-views and scenarios isn't supposed to give us a means to forecast *the* future, all we are after is some way of looking at the significant factors which affect the way the world is going, so that the problems which need to be solved most urgently get the attention they deserve. And that brings us squarely back to the problem of war, using this rather unpleasant prospect as an introduction to the three different ways of looking at the future – the three different shades of spectacles – that are going to colour the discussion throughout the rest of this book.

As we have already seen, armament is not just a physical capacity to do violence but represents a great consumption of resources even if not being used. Both these aspects of the hungry war machine need to be taken account of, and are perceived in different terms in the three worldviews. To the conservatives, an ideal capitalist world would have no need of war since goods, money, people and ideas would move about freely in response to 'market' forces. To the reformist, conflict can arise from many factors, for example competition between nations for limited resources or within nations for a bigger share of the national cake – the conflict is due to imperfections and irrationalities in the system and can be resolved through care and maintenance. To the radical, however, the system is so far gone that care and maintenance are useless, and it is better to scrap everything and start again. Capitalism now is seen as synonymous with conflict – the oppression of the poor by the rich, and the struggle of the poor against oppression.

All agree, however (even the conservatives, forced to acknowledge the existence of conflict in the real world), that a low-growth inegalitarian world must be one dominated by conflict – war, civil war, or armed suppression of the have-nots.* Equally, all agree that the best prospect of avoiding war lies at the other end of the scale, in a high growth, egalitarian world of relative harmony. And that is where our interest focuses.

Technological Overkill: A Discontinuous Future?

None of the three worldviews sees continued preparation for war as inevitable, but equally none of them sees disarmament as inevitable. All, by implication, contain the prospect of doom – the ultimate discontinuity; and all contain, as well, some hint of a prospect of a happier future with a more preferable discontinuity, disarmament, of which more later. It's the need to contemplate discontinuities that makes the militaristic side of the futures

---

* What they do not agree on is whether slow growth in an inegalitarian world promotes conflict pure and simple, or whether lavish spending on war materials leads to low growth and an unequal division of the available resources. But happily the differences of interpretation make no difference to our commonsense contention that such a future is the last thing we want if it can possibly be avoided.

game different from anything in the past, and it is the prospect of technological overkill that lies at the heart of this problem.

The point is that the destructiveness of all-out war today may be so great that there could no longer be any 'winner' in the old sense – victory or defeat are equally unpleasant prospects. This is quite different from the older view, raised from time to time throughout history, that warfare might be too ghastly to contemplate. It's not just that the war itself might be ghastly beyond all previous experience, but that nothing worthwhile might be left afterwards. So the planning of military strategists, taking account of this ultimate prospect, includes the possibility not of defeat of one side by another but of the ultimate suicidal last fling of the side who thought itself to be losing bringing, in the jargon term, 'mutually assured destruction' (it is no coincidence that the acronym for this is MAD).

How has this state of affairs been reached? One approach used by military historians is to compare the lethality of successive generations of weapons over the past centuries, from which the steps shown in Figure 9 are used to indicate 'progress'. The numbers that actually go in to such a table depend on your favourite assumptions about how to define the 'lethality index', but it is pretty clear that a longbow is a more lethal weapon than a sword, that a fighter-bomber has more destructive power than a tank, and so on. And, after all, by the time we get up to the present day what does it really matter whether the 'lethality index' of a fission bomb is 49 million, 50 million, or even 100 million?

Among the new weapons which were claimed at the time of their introduction to render further wars 'impossible' were the new artillery weapons of the eighteenth century, the machine-gun, bomber aircraft and nuclear weapons. History shows that these claims were wishful thinking, in all except (so far) the last case. But history also shows that the claims may not have been *entirely* wishful thinking.

More lethal weapons haven't yet rendered war im-

59

Figure 9: A comparison of the lethalities of successive major weapons

Weapon	Lethality index*
Sword	20
Javelin	18
Bow and arrow	20
Longbow	34
Crossbow	32
Arquebus, 16th century	10
Musket, 17th century	19
Flintlock, 18th century	47
Rifle, Minié bullet, mid-19th century	150
Rifle, breechloading, late 19th century	230
Rifle, magazine, World War I	780
Machine gun (MG), World War I	13 000
Machine gun (MG), World War II	18 000
Tank, WWI (Armament: two machine guns)	68 000
Tank, WWII (one 3-inch gun, one machine gun)	2 200 000
Field gun, 16th century, ca. 12-lb cannon-ball	43
Field gun, 17th century, ca. 12-lb cannon-ball	230
Field gun, 18th century, Gribeauval, 12-lb shell	4 000
Field gun, late 19th century, 75 mm high-explosive shell	34 000
Field gun, WWI, 155 mm HE shell	470 000
Howitzer, WWII, 155 mm HE shell with proximity fuse	660 000
Fighter-bomber, WWI (one MG, two 50-lb HE bombs)	230 000
Fighter-bomber, WWII (eight MG, two 100-lb HE bombs)	3 000 000
Ballistic missile, WWII, HE warhead (V-2)	860 000
Fission explosive, 20 Kt airburst	49 000 000†
Fusion explosive, 1Mt airburst	660 000 000†

Source: Colonel T. N. Dupuy, 'Quantification of factors related to weapon lethality', annex III-H in 'Historical Trends Related to Weapon Lethality'. US Army AD 458 760–3 (Washington, DC, 1964).

Notes (Figure 9)

* The lethality index is obtained by multiplying together several values, calculated from known or estimated performance, assigned to each of the following factors:
— Effective sustained rate of fire: largest feasible number of strikes per hour, but ignoring logistical constraints (such as ammunition supply).
— Number of potential targets per strike; a target is taken to be one man, and for comparability it is assumed that the men against whom the weapon is used are standing unprotected in the open in massed formation each man occupying 4 square feet of ground.
— Relative effect: a fraction reflecting the probability of a man affected by the weapon being incapacitated by it.
— Effective range.
— Accuracy: a fraction reflecting the probability of a strike hitting its target.

possible, but each major jump up the ladder of lethality seems to have made users of the new weaponry more cautious in how they exploit it. From the middle of the nineteenth century, international laws began to appear prohibiting or limiting, by treaty, the use of some weapons. Chemical and bacteriological warfare are the obvious examples (although the success of the ban on these weapons must be at least partly because of their impracticality in most war situations!), and in recent years treaties partly banning nuclear tests, attempting to stem nuclear proliferation, and limiting manipulation of the environment for hostile purposes have all been signed by major military powers.

The snag with all this, though, is that the development of new ways of waging war continues apace, and any formal treaties limiting the use of the new methods lag behind. Just as it used to be said that Britain always went into any new war armed with the weapons that would have won the previous war, so the international legislation on arms limitation can be said to be banning the weapons of the 'last' war.

This lag isn't just a feature of the attempts to limit war – it takes the military some time to adjust to the reality of new weapons themselves, which they do by changing the way forces are deployed and protected. All the time the system is in a constant state of flux as new weapons bring about new responses; the 'ultimate deterrent' view of nuclear weapons suggests that it really may be impossible

---

– Reliability: a fraction reflecting the probability of misfires, etc.

For the tanks and the aircraft, the index is calculated as the product of:
– Armament lethality: the sum of the lethality indices for each of the armaments carried.
– Mobility: taken as the square root of the maximum speed, in miles per hour.

† For the nuclear weapon, the lethality index is calculated without consideration of delivery means (missile or aircraft) and for blast effects. The basis for the area-of-effectiveness figures which Dupuy uses to derive the number of potential targets per strike for the nuclear explosives does not seem to be strictly comparable with that used for the other weapons. But the numbers are so large that even halving them to correct for this difference still leaves staggering values of 'lethality'.

for the military to find ways to work with them, but an alternative view is that it is just taking a rather long time to adapt and that in due course nuclear weapons will become part of the accepted pattern of war. But in many ways the reality of nuclear weapons already influences military thinking, not least since by perhaps making an effective 'pre-emptive strike' possible they could *encourage* war in certain circumstances.

Until we try the system out in a nuclear war, all of these ideas remain speculation. But the build-up of force, especially nuclear force, in Europe since the 1950s depends on the assumption in military circles that their weapons can be used effectively and that 'war as we know it' can continue with nuclear weapons being used. So just how grim a prospect would such a war be?

Probably the simplest, and most striking, way of looking at the possibilities – and the discontinuity – is to look at the statistics of casualties in past wars and compare them with the prospect ahead. If we stick to battle casualties, the rather surprising result to emerge is that in the Napoleonic Wars, the American Civil War and the two World Wars the casualty rate in terms of numbers of combatants per day of intensive combat stayed steady at 15–20 per cent, in spite of dramatic increases in the lethality of the weapons being used. The explanation for this is simply the military response to new weapons, increased dispersion and better protection for the troops, already mentioned. And according to some estimates the same level of casualties could be maintained even in a nuclear war, by much greater dispersion of forces, in particular.

But this cosy estimate ignores the obvious question – what about civilian casualties? It also, because of the lags in the system, takes no account of the latest development in military thinking, the 'neutron bomb' which is designed to produce a small blast, minimising damage to valuable property (such as factories, oil rigs, or even tanks) while producing a great blast of radiation to kill as many people (by implication, less valuable and more expendable than property) as possible as quickly as possible.

Even a small nuclear brush between the two major powers is likely to bring millions, or tens of millions, of deaths. The present balance of terror is built upon the assumption that the USA must have the capacity, in all-out nuclear war, *to kill at least a quarter of the Soviet population*; in fact, they have several times this capacity, and the Soviet Union has about an equal arsenal of nuclear weapons as that of the USA. Using the naive method of extrapolating past trends forward, with statistics covering the period from 1815 to the 1970s, some experts have suggested that each time the human population of planet Earth doubles, there are wars which kill *ten times* as many people as during the equivalent period before. The implication is that from now until the end of the twentieth century *200 million people will die in wars.*

Against this sombre statistical background, remember that we have not yet considered any new developments, either in technology or politics, to change the rules of the game. The details of these speculations we can leave aside, merely noting that they might well make things worse. For a crumb of comfort, though, let's try to lay one myth, the idea that biological weapons might provide some kind of 'poor man's deterrent' and then, inevitably, a poor man's Armageddon. The reason why bans on biological weapons have been so effective is because the weapons are both expensive and impractical. The USA gave up development of these weapons in 1969 after spending almost $1000 million to little or no avail. The trouble is that such 'weapons' have an equal impact on friend and foe alike, and there is no point in wiping out your enemy with plague if you catch it too.

But having thrown out a crumb of comfort, we must also recognise the greatest hazard of all, the one most likely to upset the world system and bring that discontinuity. Where biological weapons are overpublicised, this real prospect of doom is far less noticed, having crept up on us

Type of weapon	Specific weapon* for which index is calculated	Lethality index †
NON-NUCLEAR WEAPONS		
Assault rifle	5·56 mm M16	4 200
Light machine gun	7·62 mm M60	21 000
Medium howitzer, high-explosive shell	M109 with 155 mm M107 projectiles, HE fill	1 100 000
Shoulder-fired flame-rocket launcher	M202 with 4×M74 rockets, 66 mm rockets TPA‡ fill	1 200 000
Automatic grenade-launcher, HE/frag grenades	XM174 with 40 mm M406 grenades	1 500 000
Medium howitzer, HE/frag submunition shell	M109 with 155 mm M449 'improved conventional munitions'	1 500 000
Medium howitzer, nerve-gas shell	M109 with 155 mm M121 projectiles, GB fill	2 100 000
Fighter-bomber with napalm firebombs	Phantom with 19×BLU-1 750-lb firebombs	2 400 000
Main battle tank	M60 with a 105 mm gun, a light MG and a heavy MG	3 200 000
Guided missile, tactical, HE/frag warhead	Lance with M251 warhead (860 ×BLU-63 bomblets)	3 600 000
Fighter-bomber with 'general purpose' HE bombs	Phantom with 19×M117 750-lb bombs	9 600 000
Multiple rocket launcher, HE/frag rockets	French RAP-14 with 21×140 mm rockets	12 000 000
Guided missile, tactical, nerve-gas warhead	Lance with E27 warhead, GB fill	17 000 000
Multiple rocket launcher, nerve-gas rockets	M91 with 48×M55 rockets, GB fill	19 000 000
Fighter-bomber with TPA firebombs	Phantom with developmental TPA munitions	24 000 000
Fighter-bomber with HE/frag cluster-bombs	Phantom with 19×CBU-24 munitions (670×BLU-26)	75 000 000
Fighter-bomber with nerve-gas bombs	Phantom with 19×MC-1 750-lb bombs, GB fill	210 000 000
NUCLEAR WEAPONS		
Guided missile, tactical, 'mininuke' warhead	Lance with developmental 0·05 Kt whd, airburst	60 000 000
Guided missile, tactical, 1 Kt warhead	Lance with M234 whd, middle yield option, airburst	170 000 000
Medium howitzer, 'mininuke' shell	M109 with 155 mm projectile, 0·1 Kt, airburst	680 000 000
Guided missile, tactical, 20 Kt warhead	French Pluton with AN-52 warhead, airburst	840 000 000
Fighter-bomber with 350 Kt bomb	Phantom with one B-61 bomb at highest yield option	6 300 000 000
Guided missile, strategic, 1 Mt warhead	French submarine-launched M-20 missile	18 000 000 000
Guided missile, strategic, 25 Mt warhead	Soviet SS-18 intercontinental ballistic missile	210 000 000 000

* These are US weapons unless otherwise stated; in some cases—where official estimates are unavailable in the open literature—the calculations are based on our own estimates of weapon performance.

† Calculated as in Table 1, the indices for the nuclear weapons being reduced as explained in the footnote.

‡ TPA stands for 'thickened pyrophoric agent' (currently a formulation of triethyl aluminium), a flame-munition fill that is beginning to replace napalm, the latter being considered relatively ineffective.

Figure 10: Lethality indices of some modern weapons

more insidiously. Once again, the situation is best summed up in numbers (Figure 10), although again the exact numbers don't matter so much as their relation to one another.

What has happened over the past couple of decades, as Figure 10 emphasises, is that the lethality of 'conventional' weapons has crept up, while the potential uses of 'mini' nuclear weapons has provided a range of weapons with lethality much less than even the original fission bombs. The gap between 'conventional' and 'nuclear' war and weapons systems is just about closed already, and the distinction, in military terms, is becoming increasingly blurred. We are about to lose, except in a purely psychological sense, the distinction which has made nuclear war 'unthinkable', the crucial step from conventional weapons upwards.

In some circumstances, 'conventional' weapons are now seen as a more extreme alternative than some nuclear weapons! How will this affect the future world? Once again there emerges the possibility of a branch in the road. If conventional war becomes as horrible a prospect as nuclear war, then perhaps at last *all* war will become impossible; but as we lose the distinction between the two the prospect of escalation from one to the other and on to the most powerful nuclear weapons becomes much more plausible in militaristic scenarios. It is against this background that we should look at the alternative discontinuities which may shatter any cosy projections of the road ahead, war or disarmament. But first, to complete the scene setting with a global perspective, it seems appropriate to look at the impact of war on the environment, and the possibility, now taken seriously, of deliberate modification of the environment for 'hostile purposes'.

War on the Environment

Mention of the term 'environmental warfare' usually conjures up an immediate image of subtle schemes to trigger earthquakes and bring about destruction of cities under the guise of a natural disaster, or by causing an undersea earthquake in some remote part of the globe to set off a tidal wave which would sweep coastal regions of a chosen nation, again seemingly without any human intervention. But this isn't really what 'hostile enmod' is all about. To start with, the best way to trigger an earthquake (or volcano, for that matter) seems to be to let off a nuclear bomb in an earthquake (or volcano) zone, and that isn't the kind of thing you can do covertly! Secondly, and just as important, we are still a long way from being able to predict the exact results of such tampering with the forces of nature; like biological weapons, a tidal wave in particular might well sweep over friend as well as foe. So, chiefly because of the problem of 'targeting' manipulations involving the solid Earth, the main prospects of environmental warfare concern, directly or indirectly, disturbances of the atmosphere.

Weather modification, especially rainmaking, has already been tried out by the military in Vietnam. The aim of this was to increase monsoon rains and flood out supply trails through the jungle, but during the course of the 'experiment' the rainfall during the monsoon season never rose more than two or three inches above the average level (twenty-one inches), and this is well within the limits of normal variation from year to year. Besides, this kind of tactical use of environmental weaponry, important though it is as an indication of the growing scope of the war machine, is hardly of long-term global significance. Exotic ideas such as steering hurricanes or triggering tornadoes, even less established in practical terms, fall into the same 'local' category, while attempts to disturb the balance of polar ice cover go to the opposite extreme of causing such dramatic changes in weather systems and the circulation of the atmosphere that a resulting climatic shift might well do more harm to the nation that triggered the change than to any intended victim.

Of course, some of these possible ways to tamper with the

natural balance of the environment may well become practicable – in the eyes of the military – within the next 50–100 years. But the one possibility which seems to be taken really seriously today – the possibility of disturbing the balance of the ozone layer of the atmosphere – suffices to show how all such environmental manipulation, if attempted, may play a big part in any imminent global discontinuity.

The 'threat to the ozone layer' has become a part of ecological folklore, along with the concept of limits to growth. As with that concept, the folklore version isn't quite in line with reality, and for most purposes it is reassuring to discover that the 'ozone layer' is not a delicate film or 'soap bubble' around the Earth, which might easily be popped and destroyed, but actually a rather rugged feature of the atmosphere which has been around through thick and thin for at least sixty-five million years, and can accommodate stresses far greater than anything man can yet achieve inadvertently.

Most attention to the importance of this layer has come lately through fear that it might be disrupted by the exhausts from high-flying supersonic transport aircraft, or by the gases which used to be used so extensively in spray cans, the fluorocarbons. With SST aircraft not yet a major part of world passenger fleets, and unlikely to be so for many years, and with the propellant gases in spray cans already being changed because of the 'threat' to ozone, those influences can be ignored in the immediate future. But the ozone layer *is* important to life as we know it, and *deliberate* tampering with it is another matter altogether.

For all practical purposes, the 'ozone layer' is the same thing as the stratosphere, a region of the atmosphere extending from about 15 km altitude up to 50 km, with most ozone in a band from 20 to 35 km altitude. The layer is produced by a constant dynamic balance – ozone is continually being created by photochemical reactions stimulated by sunlight, and continually being broken

67

down by other reactions, back into ordinary two-atom molecules of oxygen (ozone, O_3, is a tri-atomic form of oxygen).

This is rather like the flow of water in a deep river. Stand by the bank, and 'new' water is constantly arriving while 'old' water flows away – but the river is always there, even if the level fluctuates as these natural balances shift slightly. The same thing happens with ozone. The concentration fluctuates from day to night, with the seasons, and over the eleven-year 'sunspot cycle' of solar activity. Overall, variations by as much as 20–30 per cent occur on timescales ranging from months to decades – much bigger natural variations than any influence caused by a few Concordes or the declining proportion of fluorocarbon-based aerosol sprays.

But why should we care about fluctuations in ozone high above our heads? Quite simply because this layer does act as a kind of sunshield, filtering out ultraviolet radiation from the Sun which could burn skin, cause cancers and damage crops if it got through unchecked to the ground below. The layer also acts as a kind of 'lid' on the weather systems below, in the bottom layer of the atmosphere, the troposphere. So its removal would also affect weather and climate.

The most extreme case, taking all the ozone away, would dramatically change conditions on the surface of the Earth, and probably make life impossible for many species. Indeed one theory about the end of the age of the dinosaurs, sixty-five million years ago, ties this in with 'stress' caused partly by a dramatic depletion of ozone, itself caused by some exotic outside stress, perhaps the explosion of a nearby star.

We don't have to go to such extremes, though, to see why some military minds are interested in the prospect of deliberate manipulation of the ozone layer. A one degree centigrade cooling might not matter to many parts of the world, but could wipe out wheat-growing in Canada, according to a US National Academy of Sciences report,

68

and that's just the kind of effect that could result from tampering with the ozone. A 70 per cent increase in ultraviolet (UV) reaching the ground would cause severe sunburn within ten minutes of exposure of human skin – and if that seems a joke, remember that sunburn is so incapacitating that in many armed forces it is regarded as a 'self-inflicted wound' and a serious punishable offence.

The main effects, however, would be on plants and animals exposed in fields and unable to take shelter. Seedlings are much more sensitive, so an ozone adjustment in the Spring could have significant impact on crop yields, food supply, and the capacity of an 'enemy' either to wage war or to compete in economic terms.

But can such adjustments of ozone concentration be made selectively over the fields of your enemy, while leaving your own fields untouched? There is speculation that this might be done, perhaps even clandestinely, by releasing suitable chemicals into the upper atmosphere, to react with ozone and shift the natural balance so that the 'river' dries up. Any 'hole' in the ozone layer produced in this way would soon fill in as ozone spread from other parts of the atmosphere, so that overall that would be a slight thinning of the layer – but nothing to worry about in a system where natural variations of 30 per cent are common. It looks almost as if this form of war on the environment could be feasible. But unpleasant though such a prospect is, far worse is the likely effect of all-out nuclear war on the ozone layer – a by product that might well ensure the collapse of any nation struggling to recover from the direct effects of the nuclear weapons.

The US NAS has reported a study showing that a war involving 10,000 megatonnes of TNT equivalent would deplete the ozone layer *overall* by as much as 70 per cent, because of the way nitrogen oxides (disruptive to ozone) are produced in the hot blast of a thermonuclear bomb and carried high into the atmosphere by the resulting rising column of hot air. Local effects could be far worse, and one particular difficulty seems not to have been taken

69

account of yet in the war preparedness stakes.

An even better way to destroy ozone would be to explode nuclear bombs in the stratosphere itself – and the anti-missile systems now deployed by both superpowers are designed to do just that! If such defences were ever used, they would ensure very great depletion of ozone just over the cities and farmlands of the defenders, where burns from UV and other effects could do most damage.

It may well be, in fact, that this natural mechanism by which the atmosphere is balanced may already provide us with something like the mythical 'Doomsday machine', since *any* all-out nuclear war, even a successful pre-emptive first strike, might result in so much damage to the ozone balance worldwide, and so much damage to crops, that it would mean the collapse of 'civilisation as we know it'. Not the end of life on Earth though – remember that the end of the reign of the dinosaurs was the beginning of the rise of the mammals. Life is pretty adaptable. But, in our terms, certainly a discontinuity sufficient to make all our forecasts invalid. So, increasingly, we see that in the not too long term we do have to face up to the prospect of discontinuity in one form or another – either war or disarmament must surely be there to change the world order some day, and this prediction can be made with far more certainty than any vision of either technological boom or Malthusian gloom.

Alternative Discontinuities: War or Disarmament?

Even if the likelihood of a discontinuity in one form or another means that ultimately 'all bets are off' in the futures forecasting game, we should still look at the possibilities. If the choice is between war or disarmament, in the long run, then any sane thinker about the future must look for future worlds in which a transition from the armed-to-the-teeth state to disarmament can be made at least reasonably smoothly. If the futures debate helps to steer the real world towards one of these paths into the future, then all futurologists would be happy to see their detailed 'predictions' fall by the wayside.

The old adage that if you want peace you must prepare for war has been around for thousands of years, and for thousands of years mankind has fought wars. The present balance of terror between East and West might seem to support the idea, since we have gone for thirty years without a major war between these opponents, while both have been armed to the teeth. But, in reality, there is every reason to believe that neither side really wants war, and welcomes the threat of war as a means to keep control over mutually agreed spheres of influence. Where the agreement between the superpowers on who should run the show breaks down (Korea, Vietnam), there is localised conflict; but where one side is agreed to hold sway no conflict occurs even under extreme provocation (Hungary, Czechoslovakia). And, of course, once local conflicts have been resolved the system can accommodate a slight shift in the boundaries between the two zones without any long-term adverse consequences overall (Vietnam, Cambodia).

Does the deterrent really deter? On the contrary, according to one line of reasoning, it encourages all-out war since, once any nuclear weapon is used, escalation will inevitably follow – and this grim line of reasoning may now be extended, as we have seen, to the use of the 'bigger' and 'better' non-nuclear weapons, as the distinctions, in military effectiveness, become ever more blurred. In an all-out war, if one side seemed to be winning, the loser would be tempted to use the nuclear arsenal in a last-ditch attempt to turn the tables; knowing this the winner would be tempted, even though winning, to use his nuclear arsenal quickly to wipe out the opponent's capacity for such retaliation. In such a situation, the possibility of a nuclear strike must be present from the very beginning, and who can doubt that the more weapons there are around the more likely they are to be used?

Finally, in looking at war itself, we must remember that, as Churchill said in the original 'balance of terror' speech in 1955, 'The deterrent does not cover the case of lunatics or dictators in the mood of Hitler when he found himself in his final dugout.' This prospect goes hand in hand with the possibility of accidental war, through faults in the early warning systems or accidents with nuclear weapons themselves. War preparedness is not, after all, the way to avoid war. Yet disarmament on a scale sufficient to remove the prospect of cataclysmic war would change the political structure of the world dramatically. Not least, it would remove the basis of the present hold which the superpowers have over lesser powers – the implication that 'you'd better join my gang to get protection from the bullies over there'.

But there is a way in which this power is already being weakened, and which may encourage, through a rather unexpected route, a shift away from the use of naked military power to dominate the world. Without war, military strength depends on the opinions held of it by outsiders. If no one takes it seriously, or if other forms of power are found to be more effective, the deterrent and threat value of military power is diminished. The classic example is the way in which the economic power of the oil-producing countries was used in the early 1970s. The USA did not respond to the 'oil weapon' by sending in the troops but paid up, in spite of a few grumbles. Military power meant little in that situation, while economic power meant everything. Such events must make even the most militaristic superpower ponder on the true value of its massive investment in weapons. The possibility then arises of a future in which mass-destruction weapons have been outmoded by a changed world order, as the superpowers accept that possession of the superweapons is at best a mixed blessing and at worst a liability – a more effective counter to the OPEC increase in oil prices would come from economic strength, strength which could be increased by diverting resources away from the war machine.

This doesn't mean that such a world would necessarily be a wonderful place to live in. Economic power struggles –

not to mention 'limited' war – may be unpleasant and wasteful of resources. But it does hint that it is possible to make the first step along a future path without the threat of destruction hanging over society.

What do our three 'worldviews' tell us about the possibility of progress along such a path? The conservatives certainly see such a future as possible – indeed, in their worldview the puzzle is why the world isn't like that already! – but don't tell us very much about how to get there. The reformists go further in suggesting that progress towards disarmament can be made by a shift in attitudes and beliefs – such attitudes and beliefs as have already been attacked by the successful use of the oil weapon – and they stress, as we do, that the likelihood of conflict is great in low growth, inegalitarian worlds. The radicals go furthest of all, suggesting that the necessary change is not just one of attitudes but an overturning of society – literally 'radical' social change to produce a situation in which disarmament was a natural consequence.

Far more important than these shades of opinion, however, is the clear vision, in all worldviews, that disarmament is both a pre-condition and a consequence of a high growth, more equal world. Surely, the single most important task now facing anyone seriously involved with the study of possible future worlds is to find a way out of the vicious circle in which resources are squandered on the war machine, reducing prospects of growth, maintaining (or widening) existing inequalities both between and within nations, and thus encouraging jealousy, fear and attitudes which foster still further development of the capacity to wage war. If we can break out of the circle, the future ahead could be a bright one indeed – and that is what the rest of this book is all about.

Part Two

Three Keys to the Future: What Might We Find There?

Chapter Three

Population and Food:
The Malthusian Myth

If there is to be any future for mankind at all, we must have food to eat. The relationship between world population and available food supply lies at the very heart of any attempt to peer into the future, and compared with this even such important problems as the supply of energy and raw materials must take second place. And this relationship is what produced the most dramatic and highly publicised curves of the limits to growth modellers, some of which were shown here in Chapter One, the doomsday forecasts which made the whole futures debate take off in the 1970s outside the cosy talking shops of the professional futures games players.

These gloomy forecasts of massive famine and widespread starvation, resulting from population outstripping food production, are generally described as 'Malthusian' or 'neo-Malthusian', and so far in this book I've gone along with the accepted label. But, just as other great prophets such as Marx, Jesus Christ or Confucius might be a little puzzled by some of the things that go on in their names today, so what Malthus really said has been distorted and misrepresented in much of the modern futures debate. It's too late to change the labels now, but the school of thought now dubbed 'neo-Malthusian' is quite different from the thoughts of poor old Thomas Malthus himself. So the time has come to clear up the confusion and sort out these differences, as a basis for tackling the puzzle of what is

77

widely known as 'the world food problem'.

The neo-Malthusian argument, the gospel as interpreted by the disciples of gloom today, contains the assumption that there is a fixed limit to the amount of food we can produce in the world, and that we are running up close to that limit – or may even have reached it – today. As population increases, it bumps up against this ceiling – sometimes called, incorrectly, a 'Malthusian limit' on population – and then collapses drastically as millions starve. The implication, generally unstated, is that this is a new problem and that this kind of limit to growth has not applied during the past history of mankind, except perhaps in very special local circumstances.

But what Malthus actually pointed out in his famous *Essay on the Principle of Population*, first published in 1798, was quite different. Having trained as a mathematician, he realised that human populations must always increase geometrically (doubling at regular intervals) unless held back by some natural system of checks and balances. Indeed, *any* living species would suffer a dramatic population explosion if allowed to reproduce to its full capacity unchecked, as we can see even more dramatically by considering the examples of fast-breeding creatures such as rabbits, rats, or insects, when they are placed in unusually favourable conditions. The basic, and rather modest, assertion which Malthus made was that the human race has the capacity to double its population every twenty-five years, a rate rather slower than that actually found to be happening at that time in America. All this requires, after all, is that each couple should produce four children by the age of twenty-five – well within human capacity!

The next step in the argument, as put forward by Malthus, is a bit more shaky but still plausible. He held that production of food simply could not increase fast enough to keep step with such a dramatic increase in human population, so that *if such an increase in population actually occurred* famine would cut back the number of people to match agricultural production. But this is just one of the checks and balances which can keep population under control, and the important idea to emerge from the original formulation of Malthus's own ideas is that:

> The natural tendency to increase is everywhere so great that it will generally be easy to account for the height at which the population is found in any country. The more difficult, as well as the more interesting, part of the inquiry is to trace the immediate causes which stop its further progress . . . What then becomes of this mighty power . . . what are the kinds of restraint, and the forms of premature death, which keep the population down to the means of subsistence.*

What Malthus is saying is that since, quite clearly, human population has *not* through recorded history doubled every twenty-five years or less then there must be natural checks – famine, war, pestilence, disease – which hold it back. Put like this, it seems a statement of the obvious.

Now, we see the Malthusian limit to growth not as something looming up ahead which we might run into, but as something which has always operated to hold population growth back. The population growth of the past century or so, the dramatic explosion of population, is seen in its proper Malthusian context as a result of removing or reducing some of these natural checks. Rather than there being a problem ahead 'when the food limit is reached', we can say that, in a sense, population expands to consume the food available. Not 'limits to growth', but 'growth to the limits'.

The distinction to be made is between the question raised by Malthus, and the answer which he provided to that question. Certainly there may be, at any time, constraints on the production of food and the goods needed to sustain a high standard of living – the good life –

---

* From the second version of Malthus's work, published in 1803, as quoted by Antony Flew in *Malthus* (Pelican, 1970).

79

but these checks are not always the same. Technological change, in particular, has shifted the position of these checks dramatically since Malthus's time, and may well do so again.

The adjustment of population in line with the active checks on it is and always has been inevitable; recognising this, it is up to us to choose the most pleasant way to adjust. And this is where Malthus envisaged problems. If we eliminate some of the natural checks, such as disease, it seems inevitable that the power of the other checks, such as famine, will increase. So Malthus argued, for example, that in the long run the reforming Poor Laws of the nineteenth century would be a bad thing, encouraging population growth which would then be checked by other causes. And this is where he got his bad name, in the view of reformers, and where his original ideas gradually became distorted into the present neo-Malthusian gloomy scenarios.

What Malthus failed to see was the dramatic effect of the application of improved technology to agriculture, and the huge increases in productivity which have fed the population explosion. The explosion *follows* from the increased ability to produce food, which changes the earlier Malthusian balance. And, clearly, this boom cannot go on indefinitely, if only because in a few hundred years it would leave us with standing room only on Planet Earth! But that does *not* mean that world famine is inevitable, since as Malthus recognised there are other checks on population which can be applied. Most simply, we can choose not to have large numbers of children, as effective contraception becomes widely available. This is surely the most pleasant alternative, the others being famine, war, or disease, and as we saw earlier it goes hand in hand with an improvement in living standards, the 'demographic transition' so beloved of the prophets of boom.

Yet again the message is clear. Some growth is needed to provide the kind of society in which people *choose* to have fewer children, and a more equitable distribution of available wealth is needed to speed the process up in the poorest parts of the world.

The Wild Extremes

How much food can we really expect to produce in the next hundred years or so? And how many people can we feed adequately? As always, the range of views is dramatic, and some forecasts from the mid-1960s have already proved wildly wrong. Even such august bodies as the US President's Science Advisory Committee were then to be found predicting world famine for the mid-1970s, a period we have now lived through with no more calamities than we have, regrettably, become used to. Now, you can find gloommongers forecasting the big crunch in the 1980s; and so it goes on. The danger grows that as more such forecasts are proved wrong, as the world food 'system' judders from crisis to crisis without quite falling apart, people may begin to feel that the problem has gone away, just when a real crunch does come. For, although it seems that there is no technical reason why we should not be able to feed a population even double that of the world today, we can only do so effectively if the machinery of the world food system is overhauled.

Just what might be practicable? Well, if you ignore the 'political' difficulties and look only at the technological possibilities it seems that the sky's the limit. Herman Kahn, speculating on likely progress from available resources and improving technology, sees a world 200 years ahead in which everyone might be fed at the level of the USA today, and seems to think that this would be a good thing. Looking ahead along one possible path into the future, his Hudson Institute team* foresees three stages in agricultural development.

First, conventional food produced by conventional means: secondly, conventional food produced by un-

* See *The Next 200 Years*.

conventional means (by today's standards): and thirdly, unconventional food produced by unconventional means. So the most dramatic forecasts for two centuries ahead involve an element of science fiction. It is, however, informed science fiction, and while it would be extravagant to expect it all to come true, it is very likely that some of these technological possibilities will become reality. Hydroponics, the technique by which food plants are grown under controlled conditions without soil; single cell protein grown from any form of organic waste, such as wood or sewage, in microbial form to feed to animals which are eaten by people; ways to make the protein present in many leafy plants edible for human beings; and other more or less exotic ideas. By 1980 20 per cent of the 'meat' in US processed meat foods was actually 'artificial' — vegetable protein, chiefly from soya.

This shows how high technology is creeping into the food system – although it does seem a peculiarly odd system that changes a nutritious, palatable soya bean into artificial meat, using a great deal of energy in the process, instead of feeding it direct to the consumer.

But let's not worry about the exotica here. For it turns out that even sticking to the most straightforward of these categories, conventional food produced by basically conventional means, and assuming no 'new' technology but only likely improvements in already tried and tested techniques, we can feed the likely – even the unlikely – population of the world in the decades ahead. Some of Kahn's more extreme ideas are bound to prove practicable, providing more scope to feed a large population. So our estimates in the rest of this chapter are really on the cautious side. Where, then, is the real 'world food problem'? As Susan George has argued in her emotional and powerful book *How the Other Half Dies* (Pelican, 1976), the root causes of starvation in the world today, and for the immediate future, are politics and poverty – not the mythical inability of the world's agricultural system to produce enough food for its population.

A Background of Confusion

The food pessimists with their tales of gloom have infiltrated the public consciousness to a far greater extent than the optimists – and this is hardly surprising since we often see in our newspapers, or watch on TV, reports of famine and crop failure, but hardly ever get any news of all the people who are now better fed than before and the crops that produce the required goods. All of this public attention rather swamps a crucial feature of the pessimistic work, which makes it seem rather like an unstable pyramid balancing on its point. Almost all of the pessimists in the limits to growth school take as their starting point a report of the US President's Science Advisory Committee (PSAC) issued in 1967, and combine this with US Department of Agriculture and UN Food and Agriculture Organisation (FAO) projections of population and demand for food.

That vital (to their argument) PSAC report has already been proved over-pessimistic by events since 1967; there is much more 'slack' in the world food system than these reports suggest, and furthermore the *potential* productivity of the world – even leaving aside the optimistic extremes – is much greater than they imply.

Studies like that of Mesarovic and Pestel (*Mankind at the Turning Point*), using a rather high estimate of the basic human nutritional requirement set at 2200–3000 kilocalories and 70 gm of protein per day and coupling this to an estimate that 95 per cent of all available land in south-east Asia is already in use while most of Africa is 'unusable', have added to the gloom. Yet, during the decades since the Second World War while there has been a widespread belief that world food supplies are already insufficient, or at best about to be overtaken by increasing population, *food production has actually been growing more rapidly than population*! Over those decades, food production worldwide has risen by an average of 2·8 per cent a year, and population by only 2 per cent a year,

83

according to official FAO figures.

Obviously, there is a real problem involving food *distribution*, since in some underdeveloped regions of the world many people are seriously undernourished. But there is no problem of food *production.*

More confusion arises at every step of the calculation used by the pessimists. Food crises are predicted by dividing the total world food supply projected for a given year by the total world population – yet neither figure is known accurately even for the present year. Even where good population censuses are available the predictions can go astray, as has been shown by the recent hurried revision of estimates downwards in Britain to try to catch up with the rapidly changing population pattern.

How much worse are projections based on very incomplete records from Africa, say? And how good are the food supply figures? The only available records of food supply are based on food which is bought and sold on the markets with other sources guessed at, and the 'guesstimates' deliberately kept *low* to avoid underestimating world food needs. Yet a very large number of the poorest people in the world eat food as soon as they harvest it, so that it never enters these records. The recorded totals must be underestimates. But, even accepting them at face value, in 1970 the *marketed* food alone was sufficient to provide for every member of the estimated world population a diet containing 2420 kilocalories per day: the official FAO/ World Health Organisation (WHO) requirement for a healthy diet, published in 1971, was 2354 kilocalories per day, and even that may be an overestimate of requirements.

Look at it another way. The FAO/WHO requirements suggest that each person needs about 500 lb of grain or its equivalent each year. On average, over the past few years, 1300 million tons (Mton) of grain have been marketed each year, enough to feed 5200 million people. The present world population is thought to be about 4000 million: the poor who starve do so not because no food is available but because the food is not available in the right place, at the right time, and at a price they can afford.

It isn't even clear that poor people in the world today would be better off if they had smaller families. The poorest of the poor live partly by growing food, partly by gathering what wild foods they can, and partly by carrying out a whole range of minor activities such as animal minding, cleaning shoes, and fetching and carrying. Menial tasks perhaps – but their only opportunity to earn money to buy food, and all dependent on having as many breadwinners in the family as possible, in order to get hold of as much food as possible.

In poor countries, where more than one third of the children may die before reaching adulthood, there are good reasons to have many children. Reducing the 'labour force' reduces the family's supply of food and produces an even worse situation. Even the FAO now accepts that people starve because they cannot buy the food that is available, *not* because there is no food to buy. And this applies both to individuals within nations and in the case of poor nations trying to obtain food on the world market. Inappropriate but well intended 'aid' can make the problem worse – in the past three decades, food production in the Indian sub-continent has more than doubled, but inequalities within that group of people have also increased, with the tragic result that while the rich eat more the very poorest eat even less than they did before.

The Tragic Inequalities

Even approximate figures are hard to come by quickly, so that our present estimates of the world food situation are based largely on what was happening up to the middle of the 1970s. But there has been no dramatic change in the way food is produced and distributed since then, so these figures give a good indication of what is going on, and where some conventional ideas about the nature of the problem go astray.

In Figure 11 the amount of food and protein available in

different parts of the world is shown for 1970, as a percentage of the then standard FAO estimate of what people need (2354 kilocalories per day). These figures make interesting reading compared with reality. For example, 'Centrally Planned Asia' – mainly China – had available only 88 per cent of the FAO 'requirement', yet many visitors, of all political persuasions, reported no evidence of mass starvation in China. On the other hand, in Brazil where the Latin American situation of a sufficiency of food even by these FAO standards applied, there were reports at that time, and since, of actual starvation. Clearly,

	Calories 1970	Protein 1970
World	101	173
Developed	121	229
Developing	96	147
Asia and E. Asia	93	141
Africa	93	141
Latin America	106	172
Near East	97	147
Centrally planned Asia	88	153

Figure 11 : Availability of food as percentage of requirements*

* Note that requirements are here defined in terms of the 1970 estimates of 2354 kilocalories per head per day – a figure since revised downwards. (FAO. World Situation in Food and Agriculture, 1970)

the standard is not a good guide to reality, and taking averages even over regions of the globe conceals severe inequalities within nations.

Since its beginning in 1947, and especially since the introduction of the 'Indicative Food Plan' in 1969, the FAO has been investigating world food needs and working to provide a stimulus for research into ways of meeting these needs, while encouraging aid for projects aimed at increasing food production. Although there have been setbacks in irrigation programmes and difficulties with fertilizer production and education of poor farmers to make best use of available resources, the amount of food produced in the world has, as we have seen, increased steadily. But the inequalities have got worse.

In 1970, as Figure 11 shows, Africa as a whole achieved 93 per cent sufficiency of food calories relative to the rather high FAO standard. But there were very great variations between countries – as much as 15 per cent for crops such as wheat. Even within Africa, the poorer nations, and the poorest people within nations, did worse. Meanwhile the developed world ran even further ahead, with a 12 per cent increase in the proportion ratio (of food available to minimum food required) in the decade of the 1960s.

One result was a massive increase in stockpiles of grain in the USA until in 1972 they reached 49 million tonnes. Prices tumbled after 1969 because of this 'glut' and this led to a deliberate policy of taking land out of production, with the USA grain acreage being cut back from 120 million to 81 million acres. In response, prices rose, as planned, and then in 1972 the Soviet harvest failed and brought the USSR on to the world market as a massive purchaser of grain. By 1974 stocks were down to 24 million tonnes, with the richer countries able to afford the increased prices while the poor could not. And then, in 1977, the screw started to turn again.

Once again, grain stockpiles were rising in response to the changed situation. So, once again, we heard proposals to take land out of production. Will the cycle for the next few years follow the pattern of the early 1970s, with the poor again being squeezed as a result? Clearly, the present market system combined with the vulnerability of the poor to market forces acts to maintain and even increase the inequalities between rich and poor. The inequalities of food consumption can only be reduced either by a change in the way food is marketed or by retaining the traditional marketplace while acting to reduce the vulnerability of the poor to the policies adopted by the rich.

Of the two possibilities, the least disruptive seems to be to increase the power of the poor in the market – in other words, to help them to become, if not rich, at least less poor. Growth, we see again, is the best solution to the real problem, the problem of inequality. But how much growth? How much food does the world really need, if the FAO standards are set too high? And how much food can the world really produce, if the gloom sayers have got their figures pessimistically low?

Myths and Legends of the World Food 'Crisis'

In the early days of attempts to understand the overall global position of food supplies, the FAO naturally was concerned not to underestimate world food requirements. Their experts assumed that an adequate diet would be the amount and quality of food needed to keep a red-blooded American male bright-eyed and bushy tailed. This came out as 3000 kilocalories, including 90 grams of protein, per day. But, of course, some people get more than the average and some less, so to play safe the experts then upped this figure by 20 per cent to provide enough for people who fell below the average. This meant that, for some countries at least, the basic food 'requirements' that were calculated were actually enough to feed *four times* the population!

Not only the quantity of the food but also its quality was wrongly estimated for many years. For most of the 1960s, there was great concern about the 'protein gap', with reports of diseases (kwashiorkor and marasmas) which are generally linked with a lack of protein. But in recent years it has been confirmed that these diseases arise, in many cases, when there is a lack of sufficient calories, regardless of what proportion of the diet is protein. If the human body is hungry enough, it starts to use up protein – even including its own heart muscle – in desperation. In many Indian villages, the symptoms of 'protein deficiency' disappeared when the villagers were provided with more calories, but no more protein. This discovery is of particular importance since, by and large, increasing protein supplies is more costly than just providing more basic calories. So many well-intended aid programmes of the past ten years have not made best use of available funds and other resources, by concentrating on the wrong end of the problem.

So estimates of what the world *needs* have, so far, been steadily revised downwards. The figure of 2354 kilocalories, widely quoted today, is still on the high side for an average, and even so it covers a whole range of needs – a point often forgotten or deliberately hidden. A girl baby less than a year old needs only 820 kilocalories a day, a growing young man of 16 needs 3500. In different countries with different climatic conditions, where the amount of heavy work being done differs considerably, the amount of food needed also varies.

Another myth – India and Bangladesh, contrary to widespread belief, are *not* the most desperately hungry countries, except for the poor regions which are most frequently visited and filmed by TV crews. In comparison with their populations, the West Indies, parts of Latin America, and Africa, have a much worse problem in that a much larger *proportion* of their population is hungry. The facts are clear: for an Indian, the average life expectancy at birth is 53 years, in parts of Africa it is less than 30 years and, to put things in perspective, the present life expectancy in India is already *greater* than that of *aristocrats* in nineteenth-century Europe.

How many people are starving in the world today? The most commonly touted 'round figure' is 500 million, yet the WHO estimates that 10 million children under five years old are chronically undernourished. These children must have brothers, sisters and parents, but at an outside estimate the implication is that perhaps 60 million older children and adults, as well as these 10 million young children, are so severely malnourished that they are using up their bodily reserves of protein and literally starving. Even a figure of 70 million starving in the world is a powerful indictment of the way food is distributed today. But don't forget that this is only 2 per cent of the total world population – a *proportion* of malnourished smaller than ever before in the history of mankind. Even with the present deficient market system real progress is being made to alleviate hunger.

The Potential for Increased Food Production

As the figures estimating world food needs have come down, so estimates of the amount of food that *could* be produced have gone up. We're not even going to look at the way-out ideas of the super-optimists, but will stick with the basic traditional methods of growing crops, on land, and still find that there could be enough for all for at least the next hundred years.

The gloomy prognostications so familiar have spread out, like ever widening ripples on a pond, from the incorrect estimates preserved in the PSAC report already mentioned. This 1967 opinion gave a maximum possible area of land available for cropping of 6600 million hectares, of which about half had actually been cultivated sometime in history and only 1400 million hectares – less than a quarter – is in use in any one year now. Almost two thirds of this estimated resource lay in Africa, Asia and South America, where, it was thought, only about one third of the available land was being harvested in any one year. Other studies in the middle and late 1960s gave similar figures in the same range, providing a basis from which towers of gloom were built. Yet, even these figures show that world food production could be doubled immediately if all available land were put to effective use – and these figures underestimate the true position, as improved understanding of the Earth through monitoring from satellites, use of the new UNESCO soil maps, and results of the 11-year International Biological Programme have now shown.

Take the example of South-east Asia. The PSAC estimates indicate that 93 per cent of available land was already in use there in the 1960s, but the improved 1970s information tells us that only 73 per cent is in use even now. A world population of 7000 million can certainly be supported at existing levels of nutrition, without anything more dramatic than expected improvements of present-day technology. If we lower our standards, it has been suggested that ten times this population could be fed –

but that takes us way beyond the immediate concern of this book.

The single most dramatic, and important, piece of work on the real potential of the world food production system was carried out in 1973 and 1974 by a team at the Dutch Wageningen Agricultural University, combining soil maps, satellite data, recent research on weather and climate and so on to determine the 'Absolute Maximum Food Production of the World'. With 'best' farming practice, but no new technology (not even desalination), they arrived at a figure of just under 50,000 million tonnes of grain equivalent a year, forty times the present world production (which averaged 1268 million tonnes in 1968–70). Even if, as in 1970, only two thirds of the land were used for grain the world could, in principle, produce twenty-five times more than it does at present.*

Of course, this maximum could never be reached in practice, since some land will be used for other purposes, and best yields are seldom achieved. But you can put in another reduction by a factor of two thirds to take account of this and still have left a capacity twenty times present production. Are the Wageningen estimates wildly unrealistic? The evidence of Figure 12 hardly suggests they are, when some high yields obtained already (admittedly, sometimes under special conditions) are even *higher* than the 'maximum' figures used by the Dutch team.

Where does the scope for improved crops lie? In the real world, a shift from individual to commercial farming has *reduced* the productivity per hectare, in spite of increased use of fertilizers. Big business agriculture, with

* A story reported in the journal *Le Monde* on 17 January 1978, highlights the flexibility inherent in the present world food system, but at present unused. Bruce Harmon, of the University of Illinois, was reported as saying that a reduction of meat consumption by 3 per cent per head in the US would free 32 million hectares of land now used for growing cattle. A mere 5 per cent of this area would suffice to grow sufficient vegetable protein, such as soya, to plug the 'gap' in diet (although this would hardly be necessary in nutritional terms), and the rest of the land released by the change in diet could provide, through timber, sugar cane and other crops, enough fuel to meet 20 per cent of US energy requirements in the mid-1970s! There is no shortage of land *per se*, the problems that exist arise through our misuse of the land that is available.

emphasis on machinery, produces less from the same amount of land than small farmers who cultivate every corner, plant rows of one crop between those of another (where the wheels of machines on a 'commercial' farm would crush them), and in general pay attention to detail.

The ideal situation would be if the enthusiasm of farmers who own and manage their own land could be combined with modernisation of techniques, to increase both productivity per hectare and productivity per man. Is this

Crop	Yield		Wageningen 'Maximum'	
Rice	26	(Philippines)	16·8	
	28·6	(India)	24·2	
	irrigated			
	11	(Kenya)	21·5	
	14·0	(Senegal)	16·9	
	16·2	(Madagascar)	17·7	
Wheat	14·5	(US)	15–18	
	5	(Netherlands)	10·5	(intercropping)
	6·2	(Finland)	7·2	

Figure 12: High yields obtained in practice

really impossible? In some countries, at least (notably China), the evidence suggests that it can be done. But what is *not* needed is the mindless application of 'western' agricultural methods in the developing countries – where, as well as the need to maximise yields, there is also a need to maximise employment in any case! In this regard, at least, the Schumacher philosophy *Small is Beautiful* makes worthwhile reading, doubly so when viewed in the light of the figures above.

But by any criterion our estimate that the world could feed more than double the present population is clearly a conservative one – other things being equal.

Will the Weather Upset the Applecart?

After war, and political dirty pool, the most likely outside influence that is sometimes regarded as destined to make other things far from equal is the threat of a global detrimental shift in climate. In my view, there is very little doubt that the world is entering a cooling phase, with climates generally returning to the pattern of the nineteenth and earlier centuries after an unusually mild, even-tempered run of decades in the middle part of the twentieth century. The most significant effect of this change is that it brings with it increased variability from year to year and from one part of the world to another.

The North American winter of 1976–7 was a classic example – unseasonable warmth in Alaska which stopped polar bears hibernating, drought in the western USA, and the most severe winter snows for a century in the east. Such effects are real and do influence agricultural productivity. But their influence is much smaller than the potential gain available – a doubling of production – if available land were utilised effectively. Where the growing realisation that climate is becoming more variable is of crucial significance is in the low level of reserve stocks – deliberately maintained at a low level to benefit the rich countries that dominate the market. One really bad year could wipe out these reserves, with all that that implies. The climatic shift is just one more powerful argument in support of sensible husbandry to maintain sufficient reserves to tide the world over the worst that is likely to happen in the next 50 to 100 years.

If the world food system were run to the benefit of all the people of the world, the kind of climatic shift we are now experiencing would be no major problem. Mismanagement of the food system, however, makes climatic influences such as the widespread droughts of the early 1970s a potential 'last straw' to break the back of the system itself.

Looking a bit further ahead, however, a climatic change induced by man's activities may very well play a big part in agriculture and the whole structure of society on the Earth in the years after about 2030 AD. The prospect is of a pronounced *warming* of the globe, caused by increased carbon dioxide in the air which acts as a blanket to keep heat in (the so-called 'greenhouse effect') coupled with the waste heat from power stations and industrial activity. Professor Will Kellogg, of the National Center for Atmospheric Research in Boulder, Colorado, sees temperatures rising by 3°C or more, compared with the present, in the years after 2050 AD. The consequences are interesting, to say the least.

A slightly warmer Earth might well be a better place for mankind overall, since the aid to agriculture could boost the capacity to support life even beyond the figures mentioned so far. But, as the Earth warms, the pattern of rainfall distribution changes, and melting icecaps raise sea level to cause difficulties in coastal regions – not least London, New York and Holland.

Drier conditions would prevail in central North America, including Canada, with wetter weather across North Africa, down into East Africa and across the Middle East and India, together with many other changes. Ironically, the developed countries whose industries may cause this shift in weather patterns would be hard hit agriculturally as a result, while the poorer countries could receive a boost to their agricultural productivity in the form of more, and more reliable, rainfall!

Just one more example of the hazards involved in making any detailed forecasts beyond the next 100 years. Within the next 100 years, though, as this anthropogenic influence begins to counter the natural cooling trend, the indications are that there is very little the climate can do to disturb a sensibly organised world food system in which adequate stocks are maintained to cover bad years.

But part of that 'sensible organisation' must be a realisation that man's activities can, and do, change the climate, and that using land to the maximum capacity could prove hazardous, in the long run, unless account is taken of these interactions between man and environment. It won't be easy, but it can be done.

How Much Growth?

The estimates of need were too high, the estimates of potential were too low. But that doesn't mean everything in the garden is rosy. Potential is just that; to be fulfilled it requires an effort – large or small depending on where you start from. Can the countries which are on the edge of starvation now increase productivity rapidly without wrecking the environment? And can this be done in such a way that, for a change, the poorest people benefit and get enough to eat?

If we can answer these questions satisfactorily for the most vulnerable regions, then it follows that the world as a whole can get over this real world food crisis. And we have already identified the most vulnerable regions – Latin America, Africa (excluding South Africa), the Middle East, South and South-east Asia, the Caribbean, and oceanic islands. To see how much growth these critical regions need, we need also to set some target which defines a 'standard of living', for want of a better term, above which starvation and hunger can be assumed to have disappeared.

This must be, to some extent, a guess. But a widely accepted figure, which we see no reason to dispute, is that when a region reaches an income level per household equivalent to $1000 at 1970 prices then even the poorer people within that region have enough money to buy sufficient food to avoid starvation. This has been true in the past, for widely different regions of the world, and should still apply in the immediate future. And this immediately gives us some feel for the plausibility of taking the growth route out of the food crisis.

If we take the estimates of future population growth from *Mankind at the Turning Point* (Mesarovic and Pestel), which is hardly the most optimistic of futures forecasts, and look at the 'worst' case of high population growth, and if we then couple these population figures with the known overestimate that each person needs 3000 kilo-

95

calories of food per day, then the rate of growth of food supply needed if these four regions are to achieve minimum standards by the year 2000 are all between 3 per cent and 4 per cent. These are certainly high – so high that it is difficult to see them being sustained without help from the rich – but they are certainly physically possible, even in this worst possible case. So it is well worth looking in more detail at slightly more optimistic alternatives.

In terms of economic growth to reach our target level of $1000 per head per annum, the year 2000 does seem to be rather too close for comfort. For our four critical regions, we would be talking about growth rates pushing up to 8 per cent or 9 per cent, which is just unrealistic. But push the target date out only a little way and the required growth rate drops below 4 per cent, still high for sustained growth over a long period, but looking much more plausible. For sustained growth at only 1 per cent a year every part of the world would reach the target level within 300 years, but that is surely too long a timescale. Other crises would have intervened by then, and the high growth option offers the hope of stabilising the situation before disaster hits.

One reason for this is the likelihood of some version of a demographic transition, a levelling off of population, as standards of living rise. This would probably take about $1\frac{1}{2}$ generations, or 40 years, as people came to realise that more children would survive, and with the range of possible populations taken from *Mankind at the Turning Point* this means an overall population level settling down between 6000 and 10,000 million, which we have already seen is well within the true capacity of the world as far as food production goes. But the critical period is still likely to be the two decades to the end of this century, when the regions of the world that now produce and consume most of the world's food must provide more realistic help to the poor.

In recent years, various rich regions have stockpiled wheat, rice, beef, butter and sugar at one time or another; yet storage of any foodstuffs except grains in an unprocessed state is very expensive. If surplus stocks were distributed as overseas aid through a central international authority, world prices could be stabilised and a buffer created against emergencies such as climate-induced drought. But this, of course, would not necessarily please those who grow rich out of present instabilities in food markets.

A clear burden of responsibility rests with North America, Australasia and the EEC, the present and likely future food surplus regions to the year 2000. Perhaps it is not unrealistic to draw an analogy between the way out of the food crisis and the situation in wartime Britain, when both high production and equitable distribution of food were achieved by strong policies and strict legislation. Crises need crisis measures for their resolution, and although such firm control has been labelled authoritarian it seems no more so than the strict control exercised by society over crimes against property and the person. One example of the strain imposed by the present free-for-all system (which really means 'too expensive for many') indicates the need for some kind of policy measures to make the road ahead easier.

Inequalities within Nations

Redistribution of incomes *within* countries could do as much to remove starvation as economic growth and a fairer distribution *between* nations. Historically, there has been no simple relationship between economic growth and economic equality within nations, although it is broadly true that the richer nations divide their riches more equally between their citizens in one way or another. But the past pattern does give the following rule of thumb guide.

In modern times, the general experience has been that as the income per household rises from about $100–$200 (again, the figures are for 1970 values) to $200–$500 there is first a pronounced *increase* in inequality, with incomes in the top two or three fifths of the population (quintiles) racing ahead of those in the poorest quintiles. As the average household income moves through $500–$1000 this trend slows and then reverses, so that above the $1000 level the share of the bottom 20 per cent rises while that of the top 5 per cent actually falls. Sweden is a classic example of this later stage; although it is obviously naive to expect this exact pattern to apply everywhere, it has been nearly followed so often in the past that it must provide some guide to future possibilities. And the details matter much less than the strong emphasis that if by a deliberate effort the populations of developing countries could secure economic growth without increasing inequalities then the prospects would be much brighter than anything outlined above. And this, I should stress, takes no account of the possibility of truly revolutionary changes in society, the kind of changes which have produced, according to all the available evidence, greater equality in the 'socialist' countries of the world than in others at similar levels of development.

Some Policy Implications

In this chapter, the focus has been on what is possible. Whether or not a well-fed world becomes a reality rests largely in the hands of the policy makers: we can solve the world food problem if we really wish to. Even taking pessimistic estimates of population growth, overestimates of the food required per person, and overestimates of the income levels needed to buy that food, it turns out that growth rates of between 1 per cent and 4 per cent in the poorer regions would ensure that everyone in the world reached a reasonable standard of nutrition by 2050 AD.

World food problems are today caused not by physical constraints or 'overpopulation' but by market forces and politics, related to global inequalities. Accepting that populations will rise for the next few decades, especially in the poorest regions, and that people generally want to 'eat better', production must be increased, and adequate reserves provided against harvest fluctuations caused by climate and other factors. This increase in production depends on technical change, but technical change geared to the local requirements in the poorer regions, developing local skills and avoiding a high degree of centralisation.

Within developing countries, the poorest sections of society have in the past often had their situation made worse by development, at least in the initial phases, as shown by the situation in India. In the short term, increased demand for food by the increasingly wealthy upper quintiles reduces the amount of food available for the poor, and this now recognised problem must be taken account of in planning future development strategies. Appropriate policies might be food rationing or, probably more acceptable, the increased provision of family plots of land and the means with which families can grow their own food.

With such enlightened policies, and a fair distribution of such food surpluses as the dairy produce of the EEC, we can, after all, survive the next two decades without mass starvation. In the longer term, irrigation of land in the semi-arid zones around the tropics where growing seasons are longest, multiple cropping and planting of more than one crop in the same field and other already proven techniques seems likely to make those regions, among the present 'have-nots', the main food producers of the world by the end of the next century.

Not quite Kahnian super-optimism, perhaps, but a future world looking much more rosy than the gloomy prognostications of the neo-Malthusians. It may not be easy to achieve, but it is certainly possible: if we don't make it we will have no one to blame but ourselves, and certainly won't deserve to inherit the Earth.

Chapter Four

Energy Alternatives

In 1977, a UK Cabinet Office publication concluded that long-term energy supplies for Britain were not a problem (largely because of the assumed introduction of nuclear fast-breeder reactors) but that food supplies were likely to cause difficulties. We have already seen that both in the UK and the whole world there need not be any immediate difficulties of food supply; in this chapter we shall discover that the problems of energy requirements are, on the other hand, much greater than suggested in that and other official publications. For many years, and in spite of the 'oil crisis' of the early 1970s, the prophets of gloom have been barking up the wrong tree, and official forecasters with them. Misguided breast-beating about the food 'problem' has diverted attention from the much more real energy 'problem' – a problem which, however, can still be solved within the framework of our desired high growth/ more equal world.

There is no cause for complacency. In the view of the SPRU team, the problem of finding an appropriate energy policy for the world, and acting upon it, is the biggest problem which needs to be overcome in order to achieve the best possible future. After all, both intensive agriculture and extraction of raw materials from the earth need large energy inputs to work. And that makes it all the more important that energy should be recognised as, at present, the nearest thing to a 'Malthusian constraint' that our society must confront.

The comparison between food and energy is striking and thought-provoking. First, unlike demand for food, demand for energy rises almost exactly in line with economic growth, at least over the range experienced by any society on Earth so far. Secondly, and of crucial importance, food is a *renewable* resource – which is why improved farming practices can, we have seen, keep up with population growth for quite a while yet. Energy sources, on the other hand, are still almost entirely non-renewable fossil fuels, which get harder to extract as time goes by and more accessible reserves are used up.

The large-scale industrialisation of the world has, so far, been powered almost entirely by fossil fuels (oil, coal and gas). Any suggestion that the rest of the world can catch up by using other techniques (nuclear, solar or wave power, for example) is at best science fiction and at worst thoughtless optimism. So the first question must be whether the growth we need to reduce inequalities can, in fact, be powered by available fossil fuel reserves. *If* we are to speculate at all about the introduction of 'new' technologies, it is surely more appropriate to look at them as an aid to sustaining high living standards *after* the crucial growth stage of the next few decades has taken place.

And, of course, while looking at the technological limits of the possible we must also bear in mind the limits of the desirable. High growth at the expense of desecration of the environment by thoughtless stripping of shale oil and tar sands, for example, is unlikely to lead to the best possible future world – any more than mindless exploitation of fast breeder reactors is going to solve all our problems in a satisfactory manner.

The problem becomes one of finding the best 'mix' of energy resources to power the growth phase which we need to produce a fairer world – hence the title of this chapter. Looking fifty or more years ahead, we cannot hope to provide 'forecasts' but instead look at the implications of several alternatives, from which it is possible to draw implications about the kinds of policy decisions that must

101

be taken very soon in order to achieve a desirable future. And, as always, the scene is best set by looking first at what has actually happened in the real world over the past few decades.

If we want to look about seventy-five years ahead, then it makes sense to get a feel for the problems involved in such crystal ball gazing by looking at a recent seventy-five-year period of high growth in energy demand. We chose 1900–74, for which reasonably complete figures are available.

Between 1900 and 1974, commercial energy consumption increased tenfold, with most of the rise occurring during the last twenty-five years.* By 1974, the relative proportions of world energy consumed in the 'Western' countries was 60 per cent, while the equivalent figures for the 'Eastern' bloc and the less developed countries were 23 per cent and 17 per cent, respectively. Over the seventy-five years, the relative importance of coal fell from 90 per cent to 32 per cent, while use of oil rose from 4 per cent to 45 per cent and natural gas from 1 per cent to 21 per cent. And, again, the switch from coal to oil and gas went most quickly during the third quarter of the twentieth century.

If these trends continued until 2050 AD, annual energy consumption would reach the equivalent of 80 thousand million tonnes of coal (80×10^9 tonnes coal equivalent, or tce). The gap between rich and poor would be maintained, with about three quarters of the world's population then using about a third of the total energy consumption. But, since the experience of the recent past shows how rapidly one energy source can be replaced by another, the overall mix of fuels being used would probably be quite different from the 1974 pattern. Even if this were feasible, however, do we want to carry on with the same kind of growth as that of the recent past?

Very high levels of energy use are not necessarily a pre-

* By 'commercial' energy consumption we exclude the basic fuels such as wood and dung which never enter the commercial market. Because this indigenous energy has been rapidly replaced by commercial energy in the developing world in recent decades, the dramatic rise in commercial consumption over the past twenty-five years is partly an artefact reflecting substitution of non-commercial sources by commercial ones. But this does not affect the rest of the discussion here.

102

requisite for even the 'Western' life style of the present rich. Comparison of West Germany and Sweden with the USA today shows that while the two European countries are as rich and industrialised as the USA, and have well-developed industries consuming large quanties of energy, they produce about twice as much Gross National Product for each unit of energy used.

Although we must always bear in mind that part of this difference is due to the lavish use of energy in US agriculture* it is clear that the greater use of energy in terms of GNP in the USA (and Canada) is a result of low prices which reduce the incentive for efficiency, coupled with the huge consumption of private cars, themselves far less efficient than those of Europe. Government policies, especially on pricing, really do have a major impact on how fuel is used. So, with this encouraging thought in mind, let's look at the different components of our possible future mix of fuels (fossil fuels, nuclear power and the rest) before puzzling over the policies we need to make the best use of what we have.

Fossil Fuel: Coal, Oil, Gas and Shale

The first and most important point to be emphasised about the fossil fuel reserves of the Earth is that they are very large indeed. We may run into difficulties with the availability of cheap oil and gas, but even in the case of oil there is no prospect of a *physical* shortage of the fuel over the crucial next 100 years: and as for coal, well it's virtually coming out of our ears. The World Energy Conference in 1974 totalled up the already identified reserves of coal, excluding peat, as a staggering 1400×10^9 tonnes, of which 600×10^9 tonnes were thought to be recoverable at a profit with 1974 technology and fuel prices. The estimated

* Fertilizers, mechanisation and so on, which consume large quantities of energy, although not necessarily in the most efficient manner in terms of *global* agricultural needs. Taking half the US fertilizer to farms in India, say, would reduce US yields only slightly while boosting Indian crops a great deal, since half of the US fertilizer used per hectare is still vastly more than that now available on third world farms. And this leaves out non-productive use of fertilizer, at high energy cost, in keeping the grass of cemetery lawns and front gardens in trim!

total coal reserves, including those yet to be discovered, can be worked out from geological considerations to be at least $10,000 \times 10^9$ tonnes, of which about half ought to be counted as 'recoverable'. (See, for example, G. Foley, *The Energy Question*, Penguin, London, 1976.)

In the same year that these estimates were being made, 1974, coal production reached $2 \cdot 5 \times 10^9$ tonnes; even with the lower figure, 600×10^9 tonnes of *already known* recoverable coal, this provides ample fuel for a very long time, and plenty of slack with which to take up some of the demand at present met by oil and other fuels. This doesn't mean that the energy problem is not real, since major, probably painful, changes in the present system will be needed to make best use of available fuel. But it is surely significant that today we find major chemical and oil companies among the keenest purchasers of coal companies and coal mining rights around the world.

Coal is, of course, difficult to transport and dirty – and, as we shall see, potentially a major influence on the global environment. But there are growing possibilities for liquefaction of coal to make 'artificial' oil (achieved by Germany when cut off from oil supplies in the 1940s, and used by South Africa today) and for gasification – perhaps underground, so cutting out the mining aspect altogether.

Compared with coal, estimates of oil reserves, and of the fraction which might be recovered, are uncertain and vary widely depending on who makes the estimates. The optimists like to point out how often in the past century predictions that oil supplies were about to be exhausted have been confounded by new discoveries; the pessimists point to the growing cost of extraction in ever harsher climates (North Sea, Alaska). To some extent the pessimists are right – with the present understanding of how the continents move about the globe, and the knowledge that the seafloor is fundamentally different from the continental crust, including continental shelf, where living organisms can lay down deposits which become oil, there is no likelihood of finding any oil deposits under the deep ocean (see my book *Our Changing Planet*, Abacus, 1979).

On the other hand, actual known reserves are likely to be rather higher than officially published figures, because of the secretiveness of both oil companies and governments about the size of reserves. In these circumstances, we can take the bottom end of a 1974 OECD estimate, 200×10^9 tonnes of recoverable oil, as a conservative indication of the true situation. And each tonne of oil is equivalent, roughly, to $1\frac{1}{2}$ tonnes of coal!

In the heady days of cheap energy so recently past, the great quantities of natural gas found in most oil-producing regions was regarded as an inconvenience, and burnt ('flared off') on the spot. Hardly surprisingly in such circumstances, estimates of how much gas might be readily available are even more uncertain than those for oil, and it doesn't make much sense to try for any better accuracy than the US Geological Survey's estimate of 'between 150×10^9 and 1500×10^9 tonnes'. Even so, that constitutes a nice bonus of clean, high calorific value fuel, especially in regions such as Europe where the gas is found on the doorstep of the consumer and pipelines for its distribution already exist.

The last two widely mentioned sources of fossil fuel are the tar sands and oil shale deposits where a very great deal – perhaps hundreds of thousands $\times 10^9$ tonnes – of 'fuel' impregnates rocky material and sandstone. These are very difficult to extract, and some estimates suggest that the amount of energy used in extracting the fuel could be more than the energy produced when that fuel is eventually burnt! So, given the wealth already outlined above, and the evidence that it is extraction and distribution that are the main problems anyway, these sticky alternatives can be ignored for the time being.

Nuclear Power – Fission and Fusion

Twenty-five years ago nuclear power seemed like a gift from the gods, offering almost inexhaustible cheap electricity. Today, it seems more like the work of the devil, offering a whole array of problems and threatening, according to some people, to destroy the world. This transformation is at the heart of the present energy debate, but before looking at the implications for our energy mix perhaps it is best to stand back and run over the physics behind nuclear power.

The nuclei of atoms are made up of neutrons and protons, bound together by a strong 'nuclear' force far more powerful than either gravity or the electric force, which would otherwise tend to break up such a nucleus due to the repulsion of the positively charged protons on one another. An extra proton which passed near an atomic nucleus, moving relatively slowly, would 'see' the long range repulsion of the electric charge and be bounced away; but a proton moving fast enough to penetrate through this repulsion can get close up in the region where the stronger, but shorter range, nuclear force dominates. What happens then depends on the kind of atom – how many protons and neutrons are already present in the nucleus.

For reasons which need not be elaborated here, atomic nuclei in the middle range of known elements are the most stable and tightly bound, with 50 or 60 nuclear particles (proton + neutrons) bunched together. Most stable of all is iron, with 56 'nucleons'. Very much larger – heavy – nuclei, under the right circumstances, can be shattered by the impact of an incoming proton or, more easily, neutron, splitting into two smaller and more stable nuclei, and releasing energy as a result of the shift into a more stable state. Very small nuclei, on the other hand, such as hydrogen and helium, may welcome new nucleons under appropriate circumstances, building up bigger and, to them, more stable nuclei (carbon, oxygen) and again releasing energy. The first process is fission; the second, which powers the stars (including our Sun), is fusion. (Anyone wondering where the very heavy elements came from in the first place will be reassured to know that when stars explode as novae or supernovae so much energy is produced by fusion of light elements that the process overshoots past the 'iron island' of maximum nuclear binding, with energy being added as more complex nuclei are built up. When we smash a uranium atom on Earth and release energy, what we get is the bottled-up energy from a stellar explosion thousands of millions of years ago.)

This description is an oversimplification. Some nuclei, such as thorium-232, first welcome an extra neutron with open arms, becoming thorium-233, then form a new proton by ejecting an electron, to become uranium-233, which can itself fission into two or more simpler nuclei. But the essence of a fission reactor is that if you put a fissile substance in a heap (or pile) then shoot some neutrons at it (not protons because of the problem of electrical repulsion) enough fission takes place to produce heat. Once started, the process can be made self-sustaining, since each fissioning atom will set loose two or more new neutrons to cannon into other nuclei – this is the basis of a 'chain reaction'. Too much of a good thing and the reaction runs away as a nuclear bomb; just enough and you have a pile simmering away nicely to boil water into steam to drive turbines and produce electricity.

Fission not only works, it is a proven method of commercial production of electricity. Once a reactor is running, the electricity produced seems very cheap since little fuel is needed. However, the capital costs involved in building a reactor are enormous, due to the need to avoid any possibility of a runaway reaction, and because once the pile is active the radiation from it makes it impossible for people to work on it directly, requiring automation and remote control techniques. But, as well as producing electricity, radioactive elements such as uranium are also used for atomic bombs. So the whole question of nuclear power is coloured by military and political factors. In addition, there is now a widespread realisation of the problems inherent in disposing of radioactive waste products from nuclear reactors, problems which were

107

scarcely considered in the 1950s.

The debate about the so-called 'fast-breeder' reactors centres on these issues. In such a reactor, uranium-238 is used as a fuel and works by absorbing a neutron to become plutonium-239, which itself fissions. This is one step more complicated than the earlier technique, using uranium-235 which itself splits under neutron bombardment, but is claimed to be potentially much more efficient in use of fuel. The price paid for this increased efficiency is a need to use such exotic substances as liquid sodium to *cool* the reactor, and the production of very long-lived radioactive waste. It seems very doubtful whether the claimed increase in efficiency justifies the risks. By now old-fashioned, but much safer, techniques used in systems such as the Canadian CANDU reactors are not only proven reliable but might themselves be used in a more efficient 'breeder' system using the thorium process outlined above. And, in any case, the whole mad rush into more 'sophisticated' bigger and better nuclear power systems seems to rest on a very shaky foundation, the fear that there will be no alternative sources of fuel within a couple of decades.

Fusion power, on the other hand, still looks as if one day it might provide unlimited clean power. The problem here, though, is when, if ever, 'one day' will come. To stick light elements together requires the creation of conditions rather like those inside a star, and after two decades of research on the problems involved the prospect still looks a distant one. 'One day', maybe, the world will run on fusion power, the power produced by sticking nuclei of hydrogen from sea water together. But not even a super-optimist would argue that fusion power has any relevance to the problems of the next fifty to seventy-five years, and that is the period that really matters. If we get through to a more equal society after that time, there will be plenty of scope then to worry about the longer-term future.

Other Sources of Power

It's something of a misnomer to label solar energy as one of the 'other sources'. Not only does all the energy in fossil fuels come originally from the Sun thanks to photosynthesis in plants (some of them then eaten by animals) which were laid down ultimately in organic residues but, as we have seen, even fission power is derived from the energy of the stars (although not, in that case, our own Sun)! Still, the direct use of solar energy has been very much an also ran in the power game historically, along with wind power (driven by the Sun's heat) and tidal and wave power (owing something to the tidal pull of the Sun, as well as that of the Moon and the Earth's rotation). Let's face it – even hydro-electric power depends on the Sun to evaporate sea water which falls as rain into highland regions. All this involves a lot of energy. A widely quoted figure is that the solar energy falling on 20 million square kilometres of desert is four hundred times present world energy production – but that masks the difficulties of converting such energy into useful form and transporting it to where it is needed.

But certainly there is great scope to use direct solar heat with simple technology to warm water and reduce other energy demands in hot countries – which, by and large, are also poor. Hydro-electric and wind power can also be exploited more than they are at present, along with wave and tidal power, although none of these is going to be a major individual contributor to world energy needs in the next century.

Perhaps one, seductively attractive, possibility should also be mentioned – using orbiting space stations to gather solar energy which might then be beamed down to the Earth. If we needed to go to such extremes, whoever had the capacity to build the orbiting power stations would have the world by the tail, since they would control energy supply. A more elitist (that is, unequal) situation is hard to imagine, and since it is our contention that a more equal world can be achieved there is no need to look at the concept in detail. It does indicate, however, the scope for developing energy resources way beyond the rather modest needs we are putting forward in the context of growth towards equality in the next 100 years or so. Like fusion power, its time may come. But not yet.

A more attractive prospect from the pages of the optimistic literature concerns extracting energy from the warm seawater of the tropical oceans, making use of the great temperature difference between surface layers and deeper water to drive generators. And the same kind of thing can be done on land, using the heat of the rocks – geothermal power. (Again, discussed in more detail in *Our Changing Planet*.)

Some idea of the potential is given, once again, by the 1974 World Energy Conference estimates – 50 *million* $\times 10^9$ tce of heat in the top 10 kilometres of rock, about a quarter of it under land. Obviously not much could be exploited, but, after all, the present known recoverable coal reserves are about one thousandth of 1 per cent of that total figure, making this a classic example of a very little of the total being able to go a very long way in human terms.

But all this is the icing on the cake. Fossil fuel, a judicious use of sound nuclear technology, and the bits and pieces where appropriate are surely enough for our immediate future needs. So where, then, is the problem?

The Real Problems

The problems are real enough: but they are not the problems so often screamed out from the headlines. Basically, there is no problem concerning the physical availability of energy over at least the next hundred years. To stress the point from the example of oil, the problem is not that there will be no oil at all, but that there will be no more cheap oil. In Figure 13 the non-renewable energy reserves

110

Figure 13 : Summary of reserves, resources and uncertainties for depletable energy resources

Unit = 10⁹ tonnes coal equivalent

Fuel	Proven & Possible Reserves	Ultimately Recoverable Resources	Uncertainties and Problems
Coal	1,000	5,600–7,700	Environmental costs of open cast mining ; cost and ease of transport and use
Oil	300–375	276–2,760	Size of recoverable resources
Natural Gas	200	150–1,500	Size of recoverable resources ; transport and storage
Shale Oil	negligible	1,200	Environmental and economic costs
Tar Sand	negligible	225	Environmental and economic costs
Sub-Total: Fossil-H'carbons	1,500–1,575	7,451–13,385	
Uranium thermal reactors	80–100	?	Size of uranium reserves, radio-active wastes
Uranium Breeder reactor	0	At least 5,000	Plutonium, proliferation, economics
Thorium Breeder reactor	0	At least 4,000	Problems with the fuel cycle
Thermonuclear Fusion	0	Virtually unlimited	Proving technical feasibility
Geothermal Wet Rock Dry Rock	12–60 0	300–1,500 50,000,000	Using low-grade heat Proving technical feasibility
Total (Excl. fusion and geothermal dry-rock)	1,600–1,700	At least 17,000–24,000	

Note: Equivalence in tonnes of coal calculated on a thermal basis for hydro-electricity

Figure 14 : Summary of potential problems and uncertainties for renewable energy resources

Unit = 10⁹ tonnes coal equivalent per year

Energy Source	Proven Potential	Exploitable Potential	Theoretical Potential	Uncertainties and Problems
Direct Solar	?	?	80,000	Technologies of storage and use : large-scale applications
Hydroelectric Power	7	?		Capital costs, alternative uses of water in less developed countries
Wind Power	?	3–60	400	Storage
Firewood, Dung, Waste fuel crops	1–2	?	40	Collection for use ; net energy gain
Tidal Power	0·05	?	10	Environmental effects ; capital costs ; limited potential, storage
TOTAL	At least 8	At least 11–69	124,000	

Note : Equivalence in tonnes of coal calculated on a thermal basis for solar energy and for firewood, dung and waste ; on a primary fuel equivalence basis for hydroelectric, wind and tidal power.

of the world, as discussed above, are summarised; in Figure 14 a similar summary of the capacity of renewable energy resources is given for comparison. As well as the evidence that there is likely to be enough energy of one kind or another, these figures, derived from intensive study of the energy situation, highlight two other important factors which must be taken into account in long-term planning.

First, coal alone accounts for more than half of proven and possible fuel reserves. This, rather than unproven nuclear or other technology, should be the basic rock on which energy policy is built.

Secondly, the figures emphasise that the practical difficulties, combined with the inherently limited potential for energy production, make tidal power, firewood, and dung and waste conversion inevitably a minor constituent of the world energy mix.

Those are the facts; from here on, opinions and value judgements must colour anyone's view of the 'best' road to the future, and what follows is very much a SPRU view. If you are more or less of a pessimist than the SPRU team, adjust the colour of the view accordingly – but remember that it must always be kept within the range of the facts summarised in Figures 13 and 14.

Some examples may indicate the degree of optimism in our scenarios. Looking at the figures for oil and gas reserves, we see that while there may be a lot more fuel yet to be discovered in these forms there may well not be. And, unlike the super-optimists such as Kahn, we do not see large-scale extraction from oil shale and tar sands as a viable proposition. So there must be, in any prudent energy policy, consideration of the desirability of conserving oil and gas so that they are used exclusively for jobs which cannot be done efficiently by other fuels. It would clearly be wrong, in our view, to build into any energy policy an assumption, based on wishful thinking, that shale and tar sands *will* provide a major contribution.

Equally, it seems naive, to say the least, to build an

113

energy policy on the assumption that nuclear energy is the solution to all our problems. It is quite possible that nuclear-generated electricity will not even prove very useful if we can make vast quantities of it, as we shall see in the next section, but at present we shouldn't even take for granted the prospect of success in technical terms. Fusion power is certainly worth continued investigation in the hope that one day the goose will start to lay golden eggs, but, like many other people today, the SPRU team doubts the wisdom of a policy based on widespread introduction of plutonium-fuelled, sodium-cooled fast breeders, on grounds of safety, nuclear proliferation and civil liberties.

With these problems in mind, the potential outlined above for solar, wind and geothermal power looks much more attractive. The scope for growth is certainly there – and the remaining technical problems seem no more difficult to solve than those of nuclear fusion or the plutonium, fast breeder. In purely commercial terms, the return on investment in any of these three 'natural' sources of energy is likely to be far greater than that from either of those two nuclear programmes. And the natural choice also offers more flexibility, which in view of the developments of the past seventy-five years must be a desirable capacity in any long-term policy. A cautious approach, making maximum use of proven systems and the vast reserves of coal while investigating solar, wind and geothermal power can both meet the needs of the next 100 years and provide flexibility to cope with the changing pattern of demand on that timescale – provided the energy is used wisely.

Using the Energy – Why Electricity?

The energy available from, say, a tonne of coal is of no practical use until it has been turned into a form appropriate for the needs of the 'consumer'. As it happens, coal is particularly versatile. It can be burned in small quantities to provide household heating; it can be converted into a form of gas for widespread distribution; and it can be burnt at power stations in large quantities, the heat being used to drive turbines and produce electricity. Without doubt, the most important energy conversion process of the twentieth century has been the generation of electricity, using a variety of fuels ranging from hydroelectricity to nuclear electricity. This made sense in the days of cheap energy – but not any more.

Electricity is easily 'transported', works, as the advertisers are so fond of telling us, at the flick of a switch, and provides for economies of scale at the point where fuel – coal, oil or whatever – is actually burnt. But the efficiency of the system in terms of what fraction of the energy in the coal actually ends up being used by the consumer is absolutely dreadful. At the start of the century, it was no better than 10 per cent – 10 tonnes of coal had to be burnt to get 1 tonne of coal equivalent energy in electrical form into the home or industry. Even by 1975, this figure had only improved to 35 per cent, and there are technical reasons why it cannot get much higher.

Not only is electricity generation wasteful in terms of fuel use, the power produced is very hard to store. You can't bottle up electricity while demand is low, to release it when demand is high, except for some very limited examples such as pumped storage with water. So, instead, we build many more power stations than we need, as far as average demand is concerned, just so that there's enough power around at the peak period on a very cold winter's day. Most of the time, a great amount of this capacity is doing the mechanical equivalent of twiddling its thumbs while waiting for a frosty day.

The electricity market today is clearly oversaturated in the developed world. Once you've got two TVs, a fridge, an audio system and an electric razor into most households demand couldn't go up much even if electricity were cheap and efficient in terms of primary fuel use. Yet 90 per cent of energy research and development is

115

geared to electricity – nuclear electricity, wave power, wind power or whatever. But you can't fly with electric power, melt iron ore by electricity or run cargo ships on batteries. If the demand for electricity did continue to rise as rapidly as in the past twenty-five years, the world would be thoroughly turned on its head by the end of this century. The fact is that, in the developed countries today, electricity demand is reaching its limit, regardless of the factors which are now making electricity more expensive and less desirable. Whatever else our best possible future world may be, it is *not* an all-electric world! And nuclear power today is good for only one thing, by and large – making electricity we don't want anyway.

Again, take one example from the heyday of the 'energy crisis' in Britain. While Government advertising on TV and elsewhere exhorted the populace to 'Save It' ('It' being energy), the nationalised electricity authority continued to run its own advertisements, on TV and elsewhere, extolling the wonders of its product and attempting to encourage increased use – in other words, 'Waste It'. Of these two campaigns, the former proved so effective (thanks also to price rises and so on) that demand for electricity fell way below forecasts and stayed there, with the result that the electricity authority decided to cancel building a huge new power station, Drax B. Now another Government department came into the ring, protesting that the new power station had to be built to provide employment – even though the building alone would add to the electricity bills of every consumer in the country, and the electricity produced from the new plant would be neither wanted nor needed.

Now, whatever else that circus may remind you of, one thing is sure. That pattern of events does not constitute any kind of energy policy, let alone the best possible one. What we would regard as sensible policy does have far-reaching implications, on employment and elsewhere. But these have to be faced in the proper manner, not by piecemeal botching up of the policy and papering over the cracks.

116

Certainly, of the two opposing views which ran into head on conflict so recently the 'Save It' line is the correct one. Caution alone should encourage us not to squander our energy reserves when we don't know quite how much is left. On the 75–100 year timescale we are looking at, there is great scope for saving energy even in building houses which are better insulated and more effectively heated, as others have stressed. But the key region of interest in terms of better efficiency must be transport, which uses 25 per cent of all energy consumed in the USA and 15 per cent in Western Europe, and is, of course, highly oil specific.

Contrary to widespread views, no radical changes are necessary even in the case of the gas-guzzling private car; cars last, on average, for ten or eleven years, and by 2030 AD we will have seen five complete 'rotations' of the car stock! A gradual saving with each new model introduced, and a gradual introduction of new technology, can do the job without any threat to the car industry and its millions of jobs.

Indeed, like it or not there is a sound case to be made here *against* extending the 'product life' in the case of cars, for fear that a rather inefficient product might then guzzle its way into the twenty-first century. In this case, at least, market forces really do work to improve efficiency as oil price rises take place, and the same is true of aircraft. Evolution, not revolution, is the way to get rid of the gas guzzlers – but a little legislation could help.

That said, however, the world of the immediate future is hardly likely to provide for lavish use of personal transport everywhere on the scale now practised in the USA. First priority in any conceivable future world must go to the use of energy to convert materials from one form to another, and Foley (*The Energy Question*) has estimated that if American cars were built today to the fuel consumption standards of European vehicles there would be an immediate saving of 80 million tonnes of oil a year – 120 mtce. Those figures just cannot be ignored.

Eliminating that kind of obvious waste from the present

117

energy system gives what the SPRU team calls a 'low conservation' profile, an annual rate of improvement in efficiency of energy use of 0·8 per cent until the year 2050. For comparison, the team has also considered a 'high conservation' profile, an annual rate of improvement of 1·8 per cent which could only be achieved by major changes in planning, public transport and pricing policy, starting out from the present situation in Western Europe. Within our 'high growth/more equal' future world, these profiles have to be set alongside others which have been widely discussed – the 'energy to burn' option where everyone ends up at the level which it was being forecast (in the early 1970s) the USA would reach by the year 2000, 28 tce per head per year; the 'USA 1970' profile where everyone reaches the standard of the USA at that time, 12 tce per person per year; and the 'Sweden 1970' profile, a humble 6 tce per person per year, which doesn't, after all, seem to have done the Swedes too much harm in spite of their cold climate. The result is surprising in view of most of the publicity that has attended the energy crisis. Discarding the super-optimism at one extreme, and finding the extreme pessimists too gloomy by half (almost exactly half), the middle path leads to a reasonably pleasant range of future worlds within our desired overall scenario. In energy terms, the end of the world is very far from being at hand – unless as a result of our own stupidity.

Three Profiles for the Future: A Flexible Mix

Any profiles used to describe the way in which high growth can produce a more equal world depend on the size of the world population and how quickly it stabilises as we come out of the 'S' curve; on the time it takes for poor countries (defined here as the 1974 poor) to industrialise; and on how much energy is needed to achieve and sustain the modernisation which, correctly approached, is a key to equality. As before, the SPRU team takes the population figures derived from the study by Mesarovic and Pestel, implying a world population between 5·2 thousand million and 6·7 thousand million by 2000 AD and a stabilised population level of between 6·3 thousand million and 11·2 thousand million by 2050. These figures, even the largest, are within the limits which can be fed adequately by the world we live in, so we are now justified in looking at the implications of such population growth in energy terms. If we could not feed such populations, of course, there would be no point in bothering about energy implications, and that is why the food question was dealt with in the previous chapter.

With this population growth, the SPRU team takes the perhaps optimistic assumption that the '1974 poor countries' will be industrialised by 2050 AD. Since individual countries have achieved the equivalent transformation in the past seventy-five years, and since the rate at which the transformation goes seems to increase with each country that industrialises, as we saw earlier, this is a reasonable goal in our drive to achieve a fairer society.

The really tricky problem, however, is deciding how much energy is needed to power this drive for equality. Five alternatives, mentioned above, are spelled out in a little more detail in Figure 15; of these, the 'energy to burn' and 'USA 1970' profiles are the extravagant alternatives, and will not be considered further here since they imply great pressure on available resources, plus significant improvement in technology, if they are to be achieved. That is not to say they are impossible, but as the presence of civilisation in Europe today shows they are certainly unnecessary and therefore wasteful. Such profiles would suggest future worlds in which problems of oil supply begin to crop up by 2000, as a comparison with Figure 13 shows, and by 2050 there would be problems even in finding enough fissile nuclear material to power reactors. In a nutshell, such worlds *depend* on the assumption that by 2050 either the nuclear fusion problem will be solved or the world will run on the power of breeder reactors.

At the other extreme, the 'high conservation' profile certainly offers a chance to eke out the available reserves, and this is the kind of profile favoured by the 'small is

119

Figure 15: Future energy demand: A choice of profiles

Unit = 10^9 tonnes coal equivalent

	The Year 2000			The Year 2050		
	Annual Consumption	% in 1974 poor countries	Cumulative Consumption 1974–2000	Annual Consumption	% in 1974 poor countries	Cumulative Consumption 1974–2050
Five High Growth, More Equal Profiles						
Energy to burn	45–70	40	585–1000	171–326	77–85	7,000–13,000
USA, 1970	25–40	40	390– 540	74–142	77–85	3,000– 6,000
Sweden, 1970	18–25	40	330– 390	37– 71	77–85	1,500– 3,000
Low conservation	15–22	40	290– 370	20– 39	77–85	1,200– 1,800
High conservation	11–15	40	250– 290	10– 19	77–85	700– 1,200

Note: Cumulative energy consumptions were calculated as the areas under logistic curves. The areas were calculated both by assuming a straight line growth of energy consumption between 1975 and 2050, and by measuring the area beneath a smooth curve on semi-logarithmic paper. The latter method was also used to calculate annual energy consumption between 1974 and 2000.

beautiful' school of thought. But it is hard to see how, in such a world, enough energy could be directed into economic growth to produce the equality we are seeking. The energy problems may be less, in terms of long-term availability of reserves, but the political problems now loom forbiddingly large.

This leaves us two profiles – or, really, one profile dressed in two slightly different suits of clothing. The 'low conservation' and 'Sweden 1970' profiles, or some mixture of the two, provide enough energy for growth in the developing world without depleting our reserves so rapidly that there is no hope of sustaining the standard of living arrived at in 2050. And they immediately show that simply in terms of energy needs alone the rush towards nuclear power has already gone too far.

Up to 1974, some 70,000 MegaWatts (MW) of nuclear electricity generating plant had been installed in the capitalist world. Following the 'oil crisis' of the early 1970s, bodies such as the OECD and UN predicted at the end of 1975 that as much as 2·5 million MW of nuclear electricity plant would be installed by 2000 AD, a rate of increase of 15 per cent per year in nuclear electricity capacity, sustained for twenty-five years.

By 1976, estimates were rapidly being revised downwards. First to 1·5 million MW by 2000, still a growth rate of 13 per cent per year, and more recently lower still. How far down must they come to meet up with real needs? Even 1·5 million MW capacity is equivalent to about 5 thousand million tce in terms of 'primary' energy inputs. Now, all that nuclear power can do is produce electricity, which is not entirely the answer to all our problems, as we have seen. So in a world using 5×10^9 tce of nuclear electricity, there would have to be about 20×10^9 tce of energy being burnt in all, coal or oil to power ships or aircraft, melt ore, drive cars and so on. And that is just for the present non-communist world – half the real world! So *if* there is to be scope for all that nuclear electricity it can only be in a world where demand for energy is up above 40×10^9 tce per year by 2000 AD – well into the 'energy to burn' sector and at least double the needs of our preferred profiles.

Within those desirable profiles, taking the same proportion of nuclear electricity to total energy needs, the amount of nuclear power required in the non-communist world by the end of this century will be well below one million MW – in round terms, somewhere between one sixth and half of present official projections. The mix we prefer is not 'nuclear all-electric', the approach which not only puts all the eggs in one basket, but in the *wrong* basket, but one with room for a bit of everything to provide flexibility.

Coal is still king in our preferred future world (Figure 16), with sensible growth in output (corresponding to the changes that took place between 1900 and 1929), giving production rising to 20×10^9 tonnes in 2050 and then levelling off. Available reserves would then last at least another 200 years.

Oil production, on the other hand, should be restrained so that it reaches no more than 6×10^9 tce per year by 2000, a quarter of total energy needs then as opposed to 45 per cent in 1974. Further slow expansion to 10×10^9 tce per year by 2050, only 14 per cent of twenty-first century energy needs, would ensure that oil had been available throughout the growth transition, and, if the optimists are right, would leave a bonus left over to lubricate the new world resulting from the transition.

Reserves of natural gas are as uncertain as those of oil, but even with the most pessimistic assumptions these too can be spun out until 2050 when annual production could reach 5 or 10 thousand million tce. As with oil, that might leave very little after the critical transition period, or it might leave a lot in reserve. What matters, again, is that this valuable resource is used wisely in the next few decades, the most critical ever faced by mankind.

Nuclear power, instead of dominating, comes in the middle of the SPRU list. Expansion at a rate of just 4 per

Figure 16: The SPRU preferred pattern of future energy supplies

Unit = 10⁹ tonnes coal equivalent

Fuel	Annual 1974	Annual 2000	Cumulative 1974–2000	Annual 2050	Cumulative 1974–2050	Further production possibilities after the year 2050
Coal	2·5	4–6	80–140	10–20	400– 850	At least for another 200 years at 2050 rate
Oil	3·6	4·5–6	100–125	5–10	325– 500	Very large uncertainty. Could last for up to 500 years or could be virtually exhausted
Natural Gas	1·7	2·8–4·6	55– 70	5–10	250– 450	Very large uncertainty. Could last for up to 200 years or could be virtually exhausted
Uranium: Thermal Reactors	0·2	1·0	13	5	110	Considerable possibilities based on further discovery of uranium, on use of thorium, and on uranium-based breeder reactors. Minimum of 9000×10⁹tce.
Uranium: Breeder Reactors	negl.	negl.	negl.	negl.	negl.	
Thorium: Breeder Reactors	negl.	negl.	negl.	negl.	negl.	
Geothermal Sources: Wet Rock	negl.	1·0	10	6	150	Between 25 and 225 years at 2050 rates
Hydroelectric Power	0·4	2	25	7	280	Indefinitely at 2050 rates
Solar Energy	negl.	1	5	2–12	70– 300	Indefinitely at much higher levels than in 2050
Wind Energy	negl.					
Natural Waste & Fuel Crops	negl.	0·5	2	1– 2	36– 50	Indefinitely at 2050 rates
TOTAL	8·4	16·8–22·1	290–390	41–72	1,620–2,690	Coal, nuclear, solar & wind resources remain considerable

Note: Methods of calculating cumulative consumption are the same as in Figure 15. In this table nuclear and hydroelectric energy has been counted in terms of equivalent primary inputs.

cent per year would provide ample opportunity to test different reactor types and settle on a safe design, while still giving 5×10^9 tce per annum by 2050. If the oil and gas were really to run dry then, there would be a secure base of proven nuclear technology on which to build for the more distant future.

The bits and pieces make up the balance – geothermal power equivalent to 6×10^9 tce in the year 2050, hydro-electricity providing 7×10^9 tce per year by then, and the smaller (but still useful) contributions from solar and wind power, fuel crops and gasification of natural waste chipping in.

Such a mix is more varied than the present (or 1974) mix of world energy sources, it provides flexibility in the sense that if there is a bit less oil than we now think the geothermal or solar capacity, say, can be increased, and by keeping a variety of options open it offers the best chance of further development beyond the year 2050, after the S-shaped transition. Two problems remain: can all this be achieved without critically upsetting the ecological balance of the Earth? And, apart from the practicalities, how can these desirable ends be achieved through political and policy means?

The Environmental Impact

The possibility of a global warming produced by the greenhouse effect of carbon dioxide added to the atmosphere by burning fossil fuels has already been mentioned in Chapter 3, and is clearly the main cause for concern in terms of the overall environmental impact of our preferred energy mix. There is a school of thought, headed by Professor Reid Bryson of the University of Wisconsin-Madison, which sees the cooling effect produced by a 'sunshield' of polluting dust particles in the atmosphere as even more important today,* but this is unlikely to be the

* See *Climates of Hunger*, by R. A. Bryson and T. J. Murray, University of Wisconsin Press, 1977.

124

chief problem in the future, even if Bryson's assessment of the present situation is correct.

First, there is a much better prospect of preventing the widespread dispersion of polluting dust particles than there is of stopping the escape of carbon dioxide from burning fuel. This is shown by the way in which big cities, for example, have become cleaner in recent years thanks to 'clean air' legislation. And, secondly, according to all the best climate models now available, any dust cooling will be overwhelmed by the size of the CO_2 warming involved in the use of fossil fuel on the scale envisaged in 'our' mix. It is increasingly clear that the net effect of mankind's activities in the twenty-first century will be to produce a global warming, and that is the situation we need to look at, briefly, here.

Where fuel of any kind is used in large quantities there are local effects on weather and climate, and any city dweller at high latitudes knows how much warmer the city is than the surrounding countryside in winter. But these purely local problems are not of any overall global significance. Different theoretical models predict slightly different detailed results of adding more CO_2 to the atmosphere, but as a US National Academy of Sciences publication *Energy and Climate* stressed in 1977 the importance lies not in the amount of difference between the models but the extent to which they agree, forecasting an increase in CO_2 of four to eight times present levels by the end of the twenty-second century. This would correspond to an increase in global temperature of about 6°C.

Now, this projection takes us as far beyond the period of interest of the present book (the next 100 years) as the end of that period is away from us now. So there is ample time for changes in society to alter the basis of the prediction by changing the energy mix, finding ways to remove CO_2 from the atmosphere, or whatever. But that doesn't mean the implications should be ignored, since it takes a long time for climate to change in response to such changes in the atmosphere. The CO_2 warming is a factor which should

125

certainly be taken into account when energy policy is being planned for the period 2000 to 2050 AD. And, in view of the implications for agriculture, in particular, discussed in Chapter 3, the question the policy makers should then be addressing is how much carbon dioxide *ought* to be allowed into the atmosphere over the twenty-first century in order to achieve an *optimum* climatic situation! This moves policy making into a whole new area of consideration, typical of the kind of problems which the world community will have to face after the great transition into a more equal world, if indeed that is achieved.

Unfortunately, none of the detailed studies of the climatic impact of energy use yet made corresponds closely to what seems to be the best energy mix. As with all new scientific studies, the investigators have looked first at the extreme possibilities – basically, either all solar and nuclear, or all fossil fuel profiles. Until more detailed work is done on different mixes, we can't say exactly how 'our' mix will affect global temperature. But one 'fossil fuel energy strategy' studied by climatologist Jill Williams and colleagues at the International Institute for Applied Systems Analysis in Austria compares roughly with our preferred profile, being based on energy consumption curves which level off at rather more than 30×10^9 tce not long after 2050 AD, and which are dominated by coal consumption until the end of the twenty-first century. The 'profile' and the 'impact' are indicated in Figures 17 and 18; even with this limited evidence it is already clear that from now on climatic considerations must be taken seriously in energy planning. The words of the NAS report mentioned above sum up the situation most aptly:

A decision that must be made fifty years from now ordinarily would not be of much social or political concern today, but the development of the scientific and technical basis for this decision will require several decades of lead time and an unprecedented effort. No energy sources alternative to fossil fuels are currently

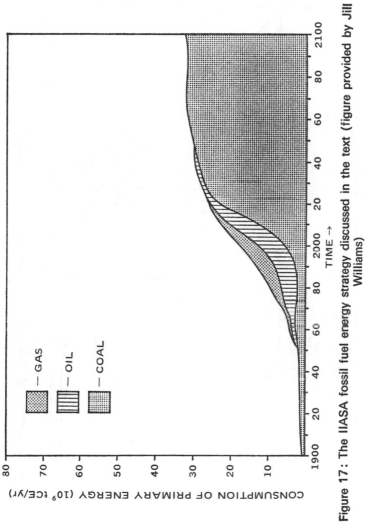

Figure 17: The IIASA fossil fuel energy strategy discussed in the text (figure provided by Jill Williams)

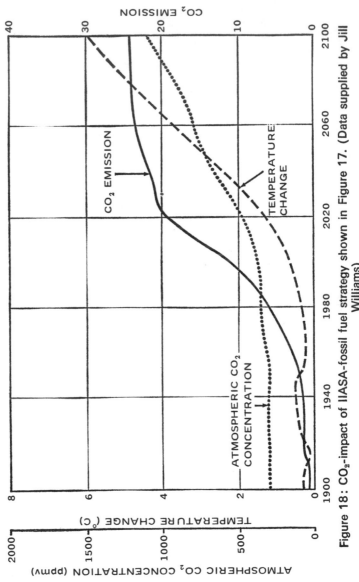

Figure 18: CO_2-impact of IIASA-fossil fuel strategy shown in Figure 17. (Data supplied by Jill Williams)

satisfactory for universal use, and, in any case, conversion to other sources would require many decades. Similarly, finding ways to make reliable estimates of the climatic changes that may result from continued use of fossil fuels could very well require decades.

Important though it is, however, to lay the ground work for those policy decisions of the twenty-first century now, it is even more important to make the right policy decisions immediately in order to steer the world along the path of more equality into the twenty-first century. We have to learn to stand up before we can decide in which direction we wish to walk.

Policies for a Flexible Response

One of the key factors in global politics today is the 'mismatch' between regions rich in energy and the regions with greatest energy demand. The classic example is the oil-rich Middle East, whose revenues are being returned in large quantities in the form of investments to the oil consumers who are the customers of the oil-rich states. A counter example is provided by a nation such as France, no longer able to dominate oil suppliers either by political or military means, turning to an ambitious nuclear energy programme in an attempt to reduce dependence on Middle East oil. But don't forget that even here the cost of that nuclear programme itself depends on the price of oil, which affects the industrial base that produces the steel and concrete that goes into building the nuclear power stations!

Figure 19 summarises the state of this mismatch as it is likely to develop until the middle of the twenty-first century. The USSR and, to a lesser extent, Eastern Europe, are well endowed with energy reserves, about forty times requirements over this period, taking a middle value of these figures. The USA has its needs covered twenty times over, China's requirements are covered about

Figure 19: Indigenous fuel supplies and energy needs, by region, 1974–2050

Unit = 10⁹ tonnes tce

Region	Ultimately Recoverable Resources				Cumulative Needs for 'Sweden 1970' Profile, 1974–2050	Uranium	
	Coal	Oil	Natural Gas	Total		Thermal	Breeder
North America	1,596–2,195	25–248	48–475	1,669–2,918	112	69	4,152
Western Europe	157– 216	11–114	7– 75 (All Europe)	175– 405[a]	84	11·7	702
Japan	5– 7	0	0	5– 7	34	negl.	6
E. Europe & USSR	3,287–4,520	24–236	39–390 (USSR only)	3,762–5,221[a]	112	?	?
Latin America	10– 14	58–578	14–137	82– 729	100–200	2·5	150
Middle East Rest of Africa	17– 24	78–779 58–576	14–139 13–130	92– 918 88– 730[b]	40– 70 70–125	} 11·5	690
China Other South East Asia	420– 578 36– 50	2– 23 17–172	} 12–115	434– 716[c] 65– 337[c]	200–300 370–700	? 1·7	? 103

Notes: (a) includes the total · or natural gas for all Europe.
(b) total for coal included in Africa rather than Middle East, since most of it is situated there.
(c) includes combined total for natural gas for China and other South East Asia.
Sources: Regional totals for coal, oil, natural gas and uranium from World Energy Conference, 1974, Figure 13.
Cumulative needs based on 6 tce per head in year 2050, and high and low population estimates of Mesarovic and Pestel.

twofold, and those of Western Europe by slightly more thanks to the huge contributions made by UK and West German reserves of coal and UK and Norwegian oil and gas.

The biggest gap between needs and available reserves is clearly in non-communist Asia, where Japan, for example, has no more than 20 per cent of its energy needs available from its own resources. By and large, the developing countries are poorly supplied with indigenous energy reserves, which makes it possible for the 'superpowers' to play a major part in influencing modernisation by providing or withholding fuel. This also raises the dangerous temptation among the have-nots to go for nuclear energy programmes as a way to get off the hook and achieve independence in energy terms. One implication is clear – if the USA and USSR really are concerned about nuclear proliferation, they should not only concentrate on coal rather than nuclear power for their own immediate energy needs, but should develop policies that ensure the provision of fossil fuels to countries that are short of them.

Although certainly possible, such scenarios are, on the face of things, decidedly optimistic. Our preferred profiles are radical by the energy-wasteful standards of the USA today, and it is hard to see how the American populace can be persuaded to become much less wasteful of oil while at the same time accepting increased coal production, not just for the home market but to help other nations. The one measure which might be effectively used to bring a major shift towards this position is a large increase in tax on energy use. Undoubtedly this would rapidly force the USA to at least match the best practice of the rest of the world in energy conservation, but equally surely it would bring the short-term (at least!) wrath of the electorate down upon the heads of the politicians courageous enough to push the legislation through. The choice is simple – do American politicians really give a damn about anything except their own chances of re-election? The evidence so

131

far suggests that, by and large, they do not.

In the USSR, of course, the situation is quite different. If the central government so decides, then the energy policies which are best for the world as a whole can be implemented with no fear of electoral disaster. And at present energy consumption per head in the USSR is less than half of that in the USA. The problems which they have to face are the technological problems of extracting the available fossil fuel, converting it into the most convenient form and transporting it widely. There is scope here for hard-nosed collaboration with technologically sophisticated, but energy-poor, countries such as Japan and the energy-poor relations of Europe, and, in addition, the propaganda value and resulting political influence of fossil fuel aid to the less developed nations makes such a course highly desirable – whatever the home market 'consumer' may think.

Whether or not such action is taken for altruistic motives, the strong government of the USSR makes it a distinct possibility. And here, perhaps, is a reason to follow our preferred path into the future which the American voter can appreciate. Without such action, the field is left clear for the developing countries to become dependent for their energy on the USSR, with all that that implies.

The present energy producers are likely to be as upset as the voters about the energy mix suggested by the SPRU team, a mix as different from the nuclear future foreseen by many today as it is from the traditional mix of the recent past. Electrical utilities, advocates of nuclear energy and manufacturers of electrical generating and transmission equipment will all be opposed to the suggested mix, along with the car industry. In today's poor countries, however, opposition is less entrenched and there is greater scope, perhaps, to get things moving in the right direction with a flexible energy mix from the outset.

Faced with such opposition, and with a relatively weak energy conservation movement, the technological problems implicit in more efficient use of energy might seem the last

132

straw to break the back of our 'best' mix. But, in fact, very little technical progress has been made since the 1950s, during the era when cheap Middle East oil discouraged any attempt to improve coal-based technology or the so-called 'unconventional' energy sources, while no one gave two hoots about efficiency in energy conservation, even in such obvious areas as home insulation. Looking at how much other technologies have changed in the past twenty-five years, there seems every reason to expect a concerted effort aimed at improving the technologies associated with fuel extraction and use to produce dramatic results. The SPRU study stresses the danger of putting all our eggs into the sodium-cooled, fast-breeder basket, when the same investment in other areas of energy research would be certain to yield large dividends very quickly. The Malthusian trap can be avoided without the Faustian bargain of widespread use of nuclear electricity; but there are powerful lobbies whose interests do not lie in encouraging the movement away from dependence on nuclear electricity and towards the flexible mix of energy sources discussed above. Here, rather than in any debate about food supplies or the population of the globe outstripping the capacity it can support, lies the ground on which the battle that will decide whether our more equal future is achieved is being, and will be, fought.

Chapter Five

The Raw Materials

Clearly, the resources of Planet Earth are finite. But, equally clearly, the actual amount of raw materials distributed through the Earth's crust – let alone its interior – is vastly more than mankind could possibly use in a millennium, let alone a mere century or so. The problem of supplying raw materials for an industrialised world is not any absolute shortage, except in a very few rare cases, but one of extracting the specific items we want from the bulk of the rock in which they are mixed, and transporting them to the places where they are required. With unlimited energy and money, these problems would disappear. But our energy and capital resources are far from unlimited, although they are adequate. So the problem becomes one of making the best use of available energy resources to extract and process the raw materials we need, at a price we can afford.

In addition, we should never forget that in extracting what we need from the Earth's crust the overall environment should not be despoiled. To some extent, these requirements are opposed to one another – ripping coal or iron ore out of an opencast mine might be very cheap in energy terms but expensive in environmental 'cost'. In view of this balance, and the sufficient but not super-abundant supply of energy discussed in the previous chapter, the *extreme* optimism about raw materials found in the pages of, say, *The Next 200 Years* is hard to justify. In the case of physical resources, however, the optimists seem to be much closer to providing an accurate vision of the future than the prophets of gloom. When we looked at the food situation, it seemed that the 'small is beautiful' approach offered a closer approximation to the SPRU consensus, because of the increased productivity which results from small farms lovingly tended by people who benefit directly from their own crops. Now, we find that the prophets of boom are not *so* wrong when it comes to the supply of raw materials for our preferred future world! This stresses the hazards of any clearcut, black and white approach to the search for a best path into the future. Small is beautiful – sometimes: boom is feasible – in the right circumstances. And it is this realisation of the shades of grey that begins to make the second generation approach to futures studies look as if it is offering real, constructive advice about how to get to a pleasant future world, not just polemic about the way the world 'ought' to have been set up in the first place, shouted from entrenched positions.

The End of the Beginning

Kahn's description of the relation between mankind's activities and utilisation of materials resources in the last quarter of the twentieth century – the end of the beginning rather than the beginning of the end – is an apt one. But to see where we are going as we come out of this initial phase it is still necessary to look back over past trends (the beginning of the beginning) and the present situation of materials use. In particular, the SPRU team has chosen to look in detail at the situation of metals, the resources which are most often alleged to be in imminent danger of running out.

As we might expect, the parallels with energy use are clear, and it makes sense to look at the present situation in terms of the rapid growth in demand for materials since 1900, and the even more rapid growth since 1950. Figure 20 sets the scene, giving pride of place to steel, of which more than 100 times as much is used as of any

Figure 20: Growth in world production of some metals, 1900-70

	1900	1925	Growth since 1900 % p.a.	1950	Growth since 1925 % p.a.	1970	Growth since 1950 % p.a.
	in million tonnes						
Steel	28	30	4·1	169	3·0	590	6·5
Aluminium (primary)	neg	0·18	–	1·51	8·9	10·3	10·1
Copper (smelter)	0·50	1·40	4·2	2·52	2·4	6·31	4·7
Lead (refined)	0·87	1·51	2·1	1·85	0·7	4·00	4·0
Zinc (smelter)	0·47	1·14	3·6	2·06	2·4	5·23	4·7
	in thousand tonnes						
Tin (smelter)	90	149	2·1	187	0·9	220	0·9
Nickel	7·6	37*	6·6	148	5·7	607	7·3

Sources: Steel from A. Sutulov, *Minerals in World Affairs* (Univ. Utah, Salt Lake City, 1972). Others from *Metal Statistics/Metallgesellschaft* (Frankfurt a.M., 1975).
* Mine production.

non-ferrous metal. The other side of the coin is shown by comparing this picture of boom with Figure 21 – as of the early 1970s, per head 'consumption' of materials in the poor countries was running at about 5 per cent to 15 per cent of the equivalent consumption by the rich.

So much for the expected. Rather less obvious to anyone who has not looked at the raw materials situation today, mining activity, as well as consumption, is concentrated in the rich, industrialised world and not in the poor countries. There are exceptions. Zaire produces 50 per cent of the world's cobalt (figures again for 1972), but Chile and Zambia produce only 11 per cent of world copper *each*, while the USA alone produces 22 per cent and the USSR 15 per cent. For iron ore, the figures are USSR 40 per cent, USA 11 per cent, the rest 49 per cent between them: for nickel, Canada 36 per cent, USSR 20 per cent: for potash, USSR 26 per cent, Canada 17 per cent, West Germany 12 per cent: and so it goes on.

All this in spite of the political and historical factors

Figure 21 : Per head apparent consumption of some materials in
different regions in 1972

Weight (kg)	OECD-rich	Rest of the world	Africa	Asia	Central and South America
Steel	566	82	27	55	
Aluminium	13	1·0	0·3	0·3	0·8
Copper	8·8	0·7	0·2	0·2	0·8
Zinc	5·8	0·6	0·2	0·2	0·5
Lead	3·8	0·5	0·01	0·01	0·5
Nickel	0·64	0·05	0·01	?	0·01
Tin	0·26	0·02	0·01	0·02	0·03

Comparative index OECD-rich = 100					
Steel	100	14	5	10	
Aluminium	100	8	2	2	6
Copper	100	8	2	2	9
Zinc	100	10	4	3	9
Lead	100	13	0·2	0·3	14
Nickel	100	8	2	?	2
Tin	100	7	3	7	10

Sources: Metal statistics from *World Metal Statistics* (World Bureau of Metal
Statistics, Birmingham, UK, monthly).
Population statistics from *UN Statistical Yearbook* (United Nations, New
York, annual).
Note: 'OECD-rich' includes all members of OECD except Greece, Spain and
Turkey.

that mean that mining itself is cheaper in, for instance,
South America. This apparent paradox is resolved by two
further pieces of information. First, since most of the production of the USA and USSR is for internal consumption,
the poor countries play a much bigger role in world *trade*
of raw materials than these figures would suggest. Secondly,
and of crucial importance to the whole futures 'debate', the
poor countries are, as far as material resources are concerned, very largely unexplored. Even after three quarters
of a century of very rapid growth in demand for raw

materials, the industrialised, rich nations still haven't had to go outside their own backyards to *look* for new supplies,* let alone get around to actually digging mines to get at the resources of the less developed countries.

This is a shocking realisation for anyone conditioned by all the propaganda of the gloommongers. But there is worse to come. We don't even need to go into details of the scope for exploration in the interior of Brazil or Eastern Siberia to see how slight our scratchings at the Earth's surface have been so far. By any reasonable standard, as far as material resources goes, the United States can be regarded as largely 'unexplored'! To quote from an official US Government Report,

> The United States is one of the few [developed] nations which has not been surveyed in sufficient detail, neither topographically nor geologically. Only parts of the country have been covered by aerial and geophysical methods which are the bases for all searches for new resources.†

In other words, the USA at least hasn't yet got around to looking properly in its own backyard.

What Others Have Said

According to the alarmist Forrester-Meadows school of thought, we are destined to run out of raw materials within a century – or, at least, to run out of any reserves that can be recovered at a 'reasonable' cost. According to the Kahnian optimists (and much of the minerals industry)

* Some of the resources might be in the backyard of a rich neighbour in any particular case, and some specific items are produced on a large scale in the developing world – but very few as we saw above.

† From *Material Needs and the Environment Today and Tomorrow*, the Final Report of the National Commission on Materials Policy, US Government Printing Office, Washington, DC, 1973. I have qualified the use of the term 'few' in this quotation in line with the context of the passage from which it is taken; of course many developing countries are still inadequately surveyed, although both for them and the USA the advent of Earth resources satellites is opening up new possibilities for locating raw materials.

there is no problem of this kind at all and the 'threat' is entirely illusory. The true picture seems to be that materials will still be available far beyond the period we are particularly interested in here, but at a cost. And the key to that cost, all too little emphasised by either of the extremist platforms, is the need for energy to extract the materials.

Apart from this key issue, the appearance of confusion among the different forecasts probably reflects the true situation. The view ahead really is confused at the level on which most futures forecasters operate, and their perspectives are almost invariably too narrow. Management in industry sees only that problems have been solved in the past, and assumes all will be plain sailing ahead; futures writers are, all too often, as ignorant of the workings of the mining industry as mining experts are of the broad sweep of the futures debate. Detailed prediction of 'the' future is impossible, yet in this area in particular almost everyone is attempting to make firm predictions, either demonstrating we are running out of resources or proving that we are not!

The SPRU team doesn't know how to make such predictions either – so, sensibly enough they don't try. Instead, their efforts have been concentrated on finding out what conditions would encourage the development of different kinds of future patterns of materials supply and use. From this approach we can now pick out the actions which are likely to bring about a situation in which growth towards a more equal world is feasible.

Perhaps the least wisely used piece of information we have – the one which has caused the most confusion – is provided by the industries' estimates of the extent of 'known reserves' of different raw materials. Almost invariably, ever since reserves started to be measured, 'known reserves' have been sufficient only for a few decades ahead (see Figure 22). This is simply a result of the way industry – even nationalised industry and government – operates. As long as enough resources are available for the next thirty years or so, there is very little need or

139

Figure 22: Known World Reserves in 1970 and Growth Rates 1950-70

	Known reserves, at present usage rate	Growth since 1950* (%p.a.)
Aluminium (Bauxite)	100 years	6·6%
Chromium	420	11·0
Copper	36	6·7
Iron	240	14
Lead	26	6·4
Manganese	97	2·0
Tin	17	2·9
Zinc	23	5·1

* Calculation includes cumulative mine production between 1950 and 1970.

incentive to invest time, effort and energy in looking for and developing new mines that are not needed. Once existing reserves start to decline, new resources are sought; and as soon as enough to last for the *next* thirty years are found the search halts once again. (In practice the process is continuous, with each year enough new resources located to maintain the necessary reserve.)

Another smokescreen of confusion is thrown up by people who worry about price rises resulting from cartels, nations who between them control some vital resource banding together to 'blackmail' the rest of the world, such as the OPEC manipulation of oil prices in the early 1970s. These are real and interesting problems, but nothing to do with how much of the raw material is actually available in the Earth's crust. A complete collapse of society resulting from a vital resource being withheld from the markets would benefit no one, and no cartel could go to such lengths – as, again, the oil example highlights. Such 'shortages' are temporary hiccups – signs of strain in our political and economic systems, but not signs that the world is 'overpopulated', 'worn out', or 'used up'.

Even Schumacher, however, the original proponent of

the 1970s 'small is beautiful' philosophy, pointed out that rather than concentrating attention on resources such as iron ore, titanium or whatever, it makes more sense to concentrate 'on the one material factor the availability of which is the precondition of all others and *which cannot be recycled* – energy'. (His italics.) (*Small is Beautiful* [Abacus edition, p. 101].)

The energy needed to produce a tonne of metal in the USA in 1973 was as little as 24 million Btu's for steel slab, and as much as 408 million Btu's for titanium. But in many cases the amount of energy actually used in extracting the raw materials from the ground – mining – is only about 15 per cent (the exact proportion, of course, depends on the ore being mined). This is the part of the production process where energy 'costs' are likely to increase. The well-established smelting and refining parts of the industry shouldn't need any more energy per tonne in the future than they do now – less, if anything, as technology improves, except when energy-intensive pollution control has to be added. For materials such as aluminium, this means that even large increases in the energy used in mining do not affect the overall energy demand of the industry very much, although, of course, the price of the energy may itself go up for reasons discussed in the previous chapter.

There are, however, materials which are extracted only at a much greater energy cost in proportion to the overall balance of the industry. Copper mining, for example, takes more than 50 per cent of the energy involved in the whole process of getting ore out of the ground and metal to the consumer, so that with unchanged technology a shift from a 1 per cent to a $\frac{1}{3}$ per cent ore doubles the *total* energy needs of the process. (3×50 per cent $= 150$ per cent for mining, plus the original 50 per cent for smelting, gives a new total of 200 per cent of the former 'energy cost'.) With this kind of very real constraint on how we can actually make use of the virtually unlimited reserves available in the Earth's crust, it becomes critically important to look at how much we actually *need* in order

141

to achieve a desirable future world. And this inversion of the problem is what makes the SPRU approach something different. The question is changed from 'how much is there?' to 'how much (or how little) do we need, and how can we best get hold of it?'

The amount we 'need' is not, in this approach, the same as the simple extrapolations of exponentially rising curves used by the extreme optimists; the importance of the approach lies in trying to establish a reasonable level of 'need', and in combining this with an attempt to assess the scope to which such demand can be varied by deliberate policy decisions.

How Much Consumption?

It makes sense to look at the possibilities in terms of one of the 'profiles' we found so useful when looking at energy supplies – the 'Sweden 1970' profile. Broadening this a little, we can say confidently that the level of consumption of materials in the OECD countries in the early 1970s is certainly more than sufficient for a satisfactory way of life, with the removal of poverty that we have seen as the key to stabilising world populations at a reasonable level, and one well within our food production capacity worldwide. To achieve such a standard, per head consumption of materials in the poor countries would have to increase tenfold; or, put another way, with no increase in population world production of materials would have to be tripled. With the reasonable assumption of a doubling of population in the developing world while this material growth was going on, we end up with a maximum 'need' of six times the present world consumption of materials per year.

But *which* materials? It doesn't follow that this profile implies six times as much steel consumption, and six times as much copper consumption, and so on. Use of some materials may grow more rapidly while others stay much the same or even decline, as materials are substituted for one another and used in different – more efficient – ways.

142

Take one example from recent history. 'Tin' cans used to be made from steel with a thin coating of tin. Over the decades, the amount of tin used per can was reduced as more efficient plating processes came into operation: today, many cans use no tin at all and are entirely made of aluminium, which among other things offers scope for improved recycling since there is no longer a mixture of metals in the can to be separated. And, a not entirely facetious example, the advent of the 'rip top' can means that there is no longer a need for steel to be used in the manufacture of special openers exclusively designed for piercing cans containing drinks!

The communications industry offers another example of the impact of substitution on the use of materials. Cable links around the world using many tonnes of copper can be replaced by satellite links (although, of course, we should take note of the energy used in placing the satellites in orbit); at a more local level, real progress is now being made with optical fibre links in which the conventional electrical impulses along copper cables are replaced by light signals sent along fine, glass fibres. Such a system can pack more information-carrying capability into the same volume as an existing cable system, while using cheap, plentiful glass instead of the copper which might be better used elsewhere. If such fibre optics systems are used to replace existing cables a great deal of scrap copper would become available – and no one has yet gone to the extreme of dredging up cables from the ocean floor to recycle the copper in them, itself a real possibility *if* the price was right.

A more dramatic example of 'substitution', strictly outside the scope of this chapter but worth looking at nevertheless, is the possibility of replacing the need for travel, in many cases, by improved communications. Such a possibility has often been hinted at in the past: the idea is simply that, in 'commonsense' terms, does it really make sense for thousands of commuters to travel into city centres to do jobs which amount to paper pushing? Wouldn't it be more sensible for the work to be sent to

143

them, using video communications links of the kind now becoming available? The savings in terms of materials and energy used in trains and cars are obvious, not to mention the improvement in working conditions for the poor commuter! The costs of the efficient communications system have to be offset against such advantages. But, once again, there is no doubt that such a change is feasible if the price is right.

As well as substitution, with cheaper, more easily available, materials replacing expensive, scarce resources, there is still a great deal of scope for increased efficiency all round. In 1957, the average age of a TV receiver when scrapped was 5·1 years: in 1971 the equivalent 'lifetime' was 10·6 years, and the sets were in any case less wasteful of materials in their construction.

Changes of this kind are not brought about through altruism or a concern for the future of mankind on the part of industry. They have occurred – and will occur – in an effort to reduce costs in cash terms. Price increases, resulting either from genuine scarcity, from the action of cartels, or through changes in taxation or other legislation are likely to have a dramatic influence on how substitution and more efficient use of materials develops in the long-term future. This is one reason why there is no hope of picking out 'the' future trend even for one metal. But it does also mean that our estimate of a 'need' for six times present world production of raw materials must be, by a long way, an upper boundary on what is 'needed'. If we make the right kind of changes, we can get away with much less than this, even for a doubled population with a Sweden (or OECD) 1970 standard of living. And that really isn't difficult to achieve, at least in technological terms.

Changes for the Better

Recycling is the ultimate means by which we might reduce the demand for virgin material. As Kahn's group put it, for the case of copper but in terms which apply much more widely, 'mining and smelting just change it from a low-grade ore below ground to a very high-grade ore above'. (*The Next 200 Years*, p. 96.) There has to be, of course, some incentive to 'mine' this high-grade ore.

At present, the incentive for the average citizen is low in the case of copper, with scrap prices kept down by the low cost of 'new' copper, about $1 a pound. But suppose the price rose dramatically – either through shortage, the actions of a cartel, or through government policy? Scrap might suddenly become so valuable, as described entertainingly by Kahn's team, that it would be worth ripping up wiring and copper plumbing and using other materials in their place wherever possible – 'the above ground reserve is scattered but easily collectable – when the price is right'.

Indeed, recycling is already used on a large scale in those many cases where the price *is* right. Around 40 per cent of world copper and lead 'production' is recycled material; in the UK alone, 55 per cent of crude steel production is 'scrap', 25 per cent of copper, and 20 per cent of aluminium. Big things, by and large, are already recycled – ships, aircraft and railway locomotives are obvious examples. The exceptions that have caught the imagination of the public and the attention of groups like Friends of the Earth are generally conspicuous in domestic refuse – beer cans and bottles, say – although not a major proportion of the materials used overall, where industry dominates. Nevertheless, there is no reason to ignore the potential of recycling domestic scrap as well as industrial scrap, although this is likely to remain a smaller contribution overall than the big time of leftover rolling stock or recycled motor cars.

Other losses of materials might also be countered – rusting is believed to 'use' 20 per cent of world production of iron and steel, and this could certainly be dealt with by protective coating techniques which are already available. Once again, the key is cost – while steel is cheap the protective coating seems expensive; if steel were to become expensive the coating would be cost effective.

So there are ways of reducing the amount of virgin raw

materials used without any comparable reduction in living standards. The overall picture is not just less gloomy than is often painted – it is a different picture altogether. The problem of meeting our demands, kept as far as possible below the six times present figure that we've set as an absolute top limit, can be tackled in two ways by a double-pronged attack – we can certainly mine more, as the prophets of boom suggest, but we can, with equal certainty and over the long term, use less, as the advocates of the simpler life proclaim.

Mining More

The top kilometre of the land surface of the Earth's crust contains enough aluminium to keep us going for at least ten thousand million years, enough iron for a hundred million years, and enough of most other metals for at least ten million years. Seawater contains, as dissolved salts, comparable amounts of raw materials, with concentrations that are already becoming commercially interesting occurring in the form of the recently discovered manganese nodules that litter some parts of the sea bed, and in the hot brine regions of geologically active sites such as the Red Sea (see *Our Changing Planet*).

Only where these concentrations have been located and proven to be suitable for profitable extraction are the rocks included in the category 'ore'. But as prices rise then automatically some rock that used to be just rock becomes ore, as lower concentrations of raw materials become potentially profitable to the mining industry. The key issue is not is it *possible* to extract the raw materials, but how much does it *cost*.

In real terms – cost expressed in terms of the cost of labour, for example – the prices of most materials do not seem to have risen since the 1920s, or even since 1900. The only obvious exception in the case of major metals is copper, which rose in price between 1960 and 1970, but has since become cheaper.

Copper also provides an example of the way 'rock' can be

146

promoted to the status 'ore'. At the beginning of this century, as high-grade ore became scarce the industry shifted away from straightforward mining and into bulk processing of rock containing a smaller percentage of copper – in effect, the industry moved into the earth-moving business. With the grade of ore halved, the amount of material to be processed is doubled, seemingly increasing costs. So far, however, these increases have been countered by innovation and improvements to existing technology. Many possibilities for further improvements are already waiting in the wings – using satellites to find ore deposits, already coming into use at one extreme, and using liquids to dissolve out the desired materials from the rocks of a 'mine' *in situ*, a more distant prospect at the other extreme, two examples among many.

The energy costs of mining more have already been discussed, and capital costs will be discussed later. The other major external factors are the amount of water used in extracting the desired material and the amount of waste produced, which inevitably increases as lower grade ores are processed.

If we include pollution under the heading 'waste', we have another factor that must increase costs, and is already claimed to have cost the US steel industry $3·5 thousand million through measures to reduce pollution, while similar measures have added perhaps 10 per cent to the cost of producing copper. (The extra cost is about 15c per pound on top of other costs.) This is just one factor which increases the capital costs of the whole industry, which is already in the biggest of big leagues. Some 70 per cent of world metal production comes from only 170 mines. The kind of investment required in such large operations needs just one thing above all else from governments and politicians – stability.

With large capital costs, the investment tied up in a mine – even if it is owned by a government – means that it must be productive for a long time. An example from the oil industry highlights the implications and difficulties that might result.

147

Companies active in the UK sector of the North Sea were affected by the decision of the incoming Labour government in 1974 to take a State stake in their operations. Uncertainty and confusion caused difficulties until the appropriate legislation was introduced and seen to be working; then, with a new *status quo* established, the job of getting the oil out and on to the market proceeded relatively smoothly once again. It isn't necessarily government participation, or high taxation, that makes life difficult for mining companies. What really causes the problems is a change in the rules of the game after the starting whistle has been blown.

But these are problems of how to make use of available resources, not problems of shortage of physical resources. I cannot resist quoting Schumacher again:

> Who could say how much of these commodities there might be in the crust of the earth; how much will be extracted, by ever more ingenious methods, before it is meaningful to talk of global exhaustion; how much might be won from the oceans; and how much might be recycled? Necessity is indeed the mother of invention, and the inventiveness of industry, marvellously supported by modern science, is unlikely to be defeated on these fronts. (*Small is Beautiful*, pp. 100–1.)

But to achieve a desirable future world we must look beyond the technological capability of industry at the social impact of more mining and greater use of resources. Above all, while it is feasible to mine more we should mine no more than we have to, and that certainly means using as little as possible to achieve 'the good life', not because there is nothing there to use but because there is more to the good life than runaway consumption of resources, however big those resources are. Too much mining would be a wasteful misuse of resources better deployed elsewhere – for example, in increasing agricultural production.

Using Less

Starting at the largest end of the scale and working down to the smallest, the first requirement to be met is that there must be countries which have resources and are prepared to develop them. In the past, deliberate decisions not to develop such resources have been few and far between: but can we assume that this will always be the case? In national policy terms, mining must be seen as profitable, either in cash terms or on some other basis, for example as an earner of foreign exchange. Mining is a good way to earn foreign currency, but it is also a good way to spend it – in the early 1970s. Peru's exports of non-fuel minerals averaged about $500 million a year, but the planned annual expenditure of two Peruvian State-owned companies for the period 1975–9 was the same, some $500 million.

With growth towards a more equal future as our aim, we cannot ignore the effects of improving the conditions of the miners themselves. Echoing the sentiments expressed by Bob Dylan, the staid pages of the *Mining Annual Review* remarked in 1975 that 'working and living conditions in most mining camps in the Andes have been a disgrace since the days of the Spaniards'. Improved technology can improve those conditions, but then the jobs themselves disappear in many cases. Is a State-owned mining company going to look with approval on, say, a machine such as a concentrator which costs $10 million but needs only 25 people to operate it – job creation at a cost of $400,000 per job?

The problem becomes one of priorities for the developing nations. Guyana, the world's fourth biggest producer of bauxite (the ore from which aluminium is obtained), employs only 6 per cent of the national workforce in mining. Is any available investment better spent on the mining industry, or in providing jobs elsewhere? And so on. The problem is exacerbated because miners tend to

149

become an elite among the working population, and in recent times have contributed directly to the downfall of governments not just in the developing world – for example, Allende's government in Chile in 1973 – but even in Britain (the Conservative government in 1974). The existence of any elite disturbs national wage policies and movement towards equality; it is easy to envisage situations in which governments might be wiser not to develop a mining industry for any or all of these reasons.

Tackling the problems by using less rather than mining more minimises the impact on the environment (there need be no great environmental 'cost' from a recycling plant), cuts down on the growth in demand for energy, and helps to stabilise the situation. Recycling of copper, zinc or lead takes only 12 per cent or 13 per cent as much energy per tonne produced as primary extraction; for aluminium the figure is as low as 3 per cent. The balance of payments may be improved in any particular consumer country by more efficient use of resources; the techniques required may be better suited than mining more, to providing jobs, with good working conditions, and so on.

The areas of waste where we can save on our use of resources have been spelled out by many writers and need not be elaborated on here. But it is important to stress that we are *not* talking about desperate belt-tightening and a dramatic change in the way of life of inhabitants of the developed countries. Perhaps the most dramatic impact might be in a cutback in the spread and use of the private car, that arch consumer of a whole variety of material resources and fuel. That doesn't mean, however, 'back to the horse' – or even 'back to the bicycle'! Michael Allaby, who leans more towards the Schumacher philosophy than that of Kahn, has suggested his version of the standard 'use less' transport future, with inter-urban passenger transport chiefly by rail but with a place still to be found for relatively efficient aircraft such as the 747 and Airbus on long-haul routes, and even private cars finding their niche in the suburbs and rural areas. (*Inventing Tomorrow*, Abacus, 1977.)

All of this picture could be brought about, over a period of twenty-five years or so, through simple policy decisions and legislation, quite painlessly and so gradually that few people would notice the change – just as few actually noticed the change in the opposite direction from 1945 to 1970. Policies which use a 'forecast' of continued growth in use of private cars as a basis for building networks of super-highways are inevitably self-fulfilling. The easier it is to use cars the more people will indeed use them. But policies designed at least to make life no easier for the motorist, plus rising fuel costs, push the traveller inexorably towards a more efficient use of available resources.

The Road to Equality

So what other kinds of policy decisions need to be made in order to ensure sensible use of available resources, with appropriate development of new mines where necessary, in order to provide the raw materials with which to build a more equal future world?

Taking recent actual growth rates as a guide (Figure 23) a future growth in consumption of materials of 5 per cent per head in the developing world seems a reasonable first guess. Such a rate of growth would bring the developing world up to our OECD (or Sweden 1970) level in about fifty years, but the demand for many materials, especially steel, would surely rise much more rapidly than our 5 per cent guideline. Just how much in the way of raw materials would actually be consumed depends on the rate at which population grows, as well as on the target set for 'sufficiency'. The range of possibilities is summarised in Figure 24 in terms of the amount of actual world consumption in 1975. And Figure 25 is included as a reminder that most of this consumption, in a world moving towards equality, must be in what is now the developing world.

It is quite clear that 'known reserves' of most materials do not match up to these large estimates of demand, so a high growth, more equal world must take up the options

151

Figure 23: Average annual growth rates in apparent per head consumption by region, from 1952/3 to 1972/3

	OECD	E. Europe	Latin America	Africa	Asia
Steel	4·2		4·9	3·0	13·0
Aluminium	6·9	8·5	13·2	18·0	12·2
Copper	2·8	5·2	3·5	3·5	8·6
Lead	1·6	4·9	2·9	1·4	8·5
Zinc	3·2	4·6	5·7	6·4	10·0
Nickel	5·7	2·1	12·4	?	38·0
Tin	0	1·3	−2·1	0	2·6

Unweighted average = 7·9% p.a.

Sources: *UN Statistical Yearbook* (various years) and *Metallgesellschaft* (various years).

discussed above – mine more and/or use less. The reduction of expectations to 50 per cent of OECD per head *consumption*, but without a comparable fall in standard of living thanks to more efficient use of materials, saves 60 per cent of the cumulative demand over the period from 1975 to 2050.

Even accepting that new sources of raw materials could be found and exploited, how much would it cost? Greatest demand would come from the region of greatest population, Asia, where by the year 2000 the SPRU team envisages as one possibility a demand for raw materials (excluding Japan) of three-quarters of the *world* demand in 1975. Fifty years later, this demand could have increased threefold: taking the costs of mining and smelting in the USA as a guide, this implies a total capital cost over seventy-five years of $650 thousand million to produce the five major materials: aluminium, copper, lead, zinc and iron. The numbers look awe-inspiring, but this is equivalent to only two years of the total GNP of Asian countries, excluding Japan, at 1975 levels. With high rates of

Figure 24 : Cumulative consumption of raw materials in all regions 1975-2050. *Units: total world consumption in 1975=1*

If consumption per head grows to:	Cumulative consumption from 1975 to	If population size is:		
		Low	Medium	High
50% of present OECD average	2000 2025 2050	39 93 150	41 100 170	43 120 210
100% of present OECD average	2000 2025 2050	43 130 230	45 140 270	47 170 350
200% of present OECD average	2000 2025 2050	62 180 380	64 190 440	6 210 570

Note : This table assumes :
(1) the per head consumption of materials in poor regions today averages 10% of that in OECD.
(2) a growth rate of 5% p.a. in per head consumption, dropping to 0% once the limit is reached; the two exceptions are for the industrialised regions : in the '50%' row, their per head consumption is assumed to fall by 1% p.a. until it is halved, and in the '100%' row, it does not change.
(3) the population projections come from Mesarovic and Pestel.

economic growth, such an investment over the next seventy-five years is certainly feasible. The numbers themselves are bound to be only a rough guide – costs may go up or down, plant that wears out must be replaced, and so on – but nothing drastic enough to rule such investment out of court is likely. Even multiplying this figure by ten, to give $6500 thousand million overall, the total cost would be only 6 per cent of Asia's GNP with growth running at 4 per cent a year.

This represents the most difficult case, along with Africa where the problems are similar. Latin America starts out from a better position than either Asia or Africa, and so has less of a problem in terms of capital investment, assuming that there really is no problem in finding new reserves.

Figure 25: Cumulative consumption of raw materials in developing countries, 1975-2050. *Units: total world consumption in 1975 = 1*

If consumption per head grows to:	Cumulative consumption from 1975 to:	If population size is:		
		Low	Medium	High
50% of present OECD average	2000 2025 2050	17 55 96	18 64 120	20 79 160
100% of present OECD average	2000 2025 2050	17 76 160	18 90 190	20 110 260
200% of present OECD average	2000 2025 2050	17 80 240	18 95 300	20 120 410

See notes to Figure 24.

For all the reasons given above, adequate new reserves are likely to be available. But it still seems essential to reduce demand – use less – if the high growth/more equal future world is to be reached in the next century. In the USA today, there are nine tonnes of steel (in buildings, machines, cars and so on) per head of population, and there is no easy way that such large quantities can be built up in the developing countries in less than a century, even if they accumulated the material at a rate of 20–30 kg per person per year, which is *higher* than the present rate in many of these countries.

The other side of the coin, however, shows that we cannot seriously expect to reduce demand so much that further mining is unnecessary. Take copper – 40 per cent of world supplies are already from recycled material, and a reduction in the need for new sources from mines, even with some improvement in efficiency, would improve substitution (using aluminium, perhaps, instead), which increases the need to mine something else!

So *both* mining more *and* using less are essential features of the road to equality. The alternative low-growth route to equality might seem superficially to be only half as difficult, removing or reducing the need to mine more. But a closer look at the route shows that it could well prove much harder.

If growth is low then there may not be sufficient economic resources available to cope with the problems of the shift to more equality. Low growth reduces the options available – it puts all our eggs in one basket, so that if anything goes wrong the kind of collapse spelled out in *The Limits to Growth* really does become inevitable.

This has important implications for the policymakers, whose actions might inadvertently set the road off down this slippery path. Proponents of the low growth/more equal view argue that effective action, perhaps by a succession of cartels, could force low growth upon the developed countries, encourage the re-distribution of resources from the rich to the poor nations, and thus allow these riches to be shared amongst the inhabitants of those poor nations. But in the real world every link in this chain is weak. *If* cartels were effective, would the rich sit back and let it happen? *If* resources were transferred to the poor nations in this way, would there be a direct benefit to the poor *within* those nations? Growth, as we have seen before, provides the opportunity for the poor to become less poor without the rich feeling threatened, and this must be a key to the successful continuation of our world beyond the next few years.

Ultimately, the choice of path must be made on social and political grounds, taking in the broad sweep of possibilities from each of our three chosen key sectors, and elsewhere. It is disturbing to realise that the social and political decisions may be founded on false information, even if made with the best motives, and the prophets of gloom have made such a noise with their protestations that modern college and school textbooks are beginning to be sprinkled with references to the world 'running out'

155

of raw materials, energy supplies and food. Strahler, for example (A. N. Strahler, *Principles of Physical Geography*, Harper & Row, 1977), cites the comment of geologist Thomas S. Lovering that:

> The total volume of workable mineral deposits is an insignificant fraction of 1 per cent of the earth's crust, and each deposit represents some geological accident in the remote past. Deposits must be mined where they occur . . . Each deposit has its limits . . . it must sooner or later be exhausted.

Contrast this with Kahn's team's equally valid comment:

> Every cubic kilometre of seawater contains approximately 37·5 million tons of solids in solution or suspension . . . an enormous amount of magnesium and varying and relatively huge quantities of gold, cobalt, lead, mercury and many other minerals. A cubic kilometre of the Earth's crust on average contains about 210 million tons of aluminium, 150 million tons of iron, 150,000 tons of chromium, 7000 tons of uranium, 80,000 tons of copper and so on. (*The Next 200 Years*, p. 104.)

Combining that with Schumacher's views on the inventiveness of industry surely provides a more secure basis for those crucial decisions, whereas blind acceptance that the end is nigh seems likely to result in decisions and forecasts that are self-fulfilling, bringing about the end of growth with catastrophic effects on the poor (at least) – effects which could have been avoided.

The time has come to see where the SPRU team stands on these issues of policy and politics. There can be enough food, energy and raw material for us to follow the path of growth into a more equal future world – but how do we make sure that we stay on the right path?

Part Three

A Choice of Paths:
How Do We Get There?

Chapter Six

Policies and Prospects: Technical and Social Change

Forecasting technical change is a dangerous business. Most people who have tried it in the past – industry, government or university academics – have made absolute howlers when attempting to spell out how much time or money will be required to develop a new technology, and how acceptable it will be to the intended users. The ghastly example of Concorde, and the continuing arguments about nuclear fast-breeder reactors, provide two obvious examples. Looking back to the mid-1960s, we can find such statements as:

The years 1965/66 will be memorable in the history of man's control of energy. Ten years ago, nuclear energy could compete with fossil power only where there was a shortage of coal and of oil or where (as in England) there was a growing reluctance to mine the thinning coal deposits, and this was possible only because the by-product plutonium could be sold to the military. It was a great surprise, even to many experts, when in these last years not only the cost per kilowatt hour of nuclear plants fell below that of coal-burning plants, but even the capital cost per installed kilowatt, by the advent of the American boiler reactor. *In Britain probably the last coal-burning plant ever to be built is now on order.* [my italics]. (Dennis Gabor, 'Material Development', in *Mankind 2000*, ed. Robert Jungk and Johan Galtung,

International Peace Research Institute, Oslo, 1969, p. 159.)

Such a statement could pass without comment at a meeting of futures 'experts' in 1967: today, it looks nothing less than lunatic. Time after time in the forecasts made in the mid-1960s we find the emphasis on the severe problems of food production, with almost a throwaway line or two to the effect that 'of course we'll have unlimited cheap nuclear power'. Gabor, for example, in the same article quoted above also said of the need for the USA to respond to the world food crisis, 'by 1975, if not before, they will have to think seriously of a real great material sacrifice in favour of the poor countries, which will involve changing their diet'.

In the face of such complete failures of past forecasts, can we hope to say anything worthwhile about technical change at all, let alone fill in some of the details of the secondary effects on society, individuals and the ecology? Scientific research by its very nature is largely concerned with the unknown and, therefore, we know that we can't predict how new technologies will emerge and whether they will be accepted with any degree of accuracy. But there are limits within which it is worth trying to work out some of the options that may be available, and to identify some of the key issues on which political decisions will have to be made in order to get to a desirable future world.

The broad direction of development implicit in the present world is clear. Massive funds go into military Research and Development, and into nuclear Research and Development, so that, inevitably, progress is going to be made in those fields, while prospects for developing new small-scale agricultural techniques or effective wind and solar power systems must be much less. The aim of the SPRU analysis is to provide guidance on the allocation of R&D priorities in order to create a more desirable future, and that is a quite different kettle of fish from trying to predict what 'will' happen if things are left to their own devices.

160

Technical Change

The question tackled by the SPRU team is crucially different from that involved in listing the way changes might occur in specific technologies. What we really need to know is whether the overall rate of technical change can be enough to produce an increase in living standards of the world's population over the next fifty years – an increase which, in round terms, must be as great as the change that took place over the past fifty years if our 'more equal' world is to become a reality. And the question can be turned on its head, to ask why should the rate of technical change now being experienced either slow down or stop? For, without such a reversal of the trend, the prospect of the amount of growth we need (though by no means any certainty that it will be appropriate) becomes inevitable.

Only then, if there is a real prospect of change continuing, do we need to know the details of the kind of change we can expect. Rather than letting things develop unguided, it becomes appropriate to try to direct the potential for change along our chosen path.

It seems that there is only one way in which beneficial technical change could be brought to a halt in the next few decades, through the occurrence of all-out war as described in Chapter Two. If we can avoid that prospect, then technical change at the level of practical applications must continue to provide a basis for growth for decades, even if we make the crazy assumption that no new scientific discoveries are ever made again. This is because it takes so long for scientific discoveries to become converted into practical applications, so that the 'pipeline' of new applications is both long and, at present, full. 'Best' techniques, for example in agriculture or energy conservation, are in use only in small areas of the world, and the large gap between 'best' and 'worst' practice today could be closed without any new discoveries or unproven techniques.

On this basis, there is scope without any new progress in

science or technology to raise the productivity of industry and agriculture in the poor nations of the world tenfold – not by wholesale and mindless application of the techniques now used by the rich, but by modifying existing technologies to fit the particular needs of each region.

So, if we imagine that there are going to be *any* genuinely new developments in science and technology, there is no doubt that a high rate of technical change and a growth in productivity that could be harnessed to pull the world along a path towards equality can continue at least well into the twenty-first century. After a hundred years or so, it might not be necessary to make further technological progress if sufficient growth of the right kind had occurred to remove the misery of the poor – *that* is when a low growth world might be an alternative. But any realistic assumption must take note of the possibility of continued scientific and technical change as far beyond our present world as we are from the late nineteenth century.

The comparison may be particularly apt. Many inhabitants of the world of nineteenth-century science were convinced that little remained to be discovered, with electricity and magnetism combined into one theory of electro-magnetism, the chemical elements all discovered and put in their places in the periodic table, and so on. This cosy picture was then upset by the discovery of radio-active decay, relativity theory and quantum mechanics. Today, once again people can be found who believe that this time we have got the world picture right, and all that is left is filling in the details. But it is easy to point to flaws in this modern picture. The nature of gravity remains unsatisfactorily explained, and attempts to combine gravity theory with quantum mechanics through studying the bizarre behaviour of matter, space and time in the vicinity of black holes offers a real prospect of progressing to a theoretical development beyond Einstein's theory of relativity, but incorporating it, just as Einstein's theory extends beyond, but incorporates, the Newtonian theory

162

of gravity. If such blue sky research seems esoteric and impracticable, it is only necessary to recall how impractical and esoteric relativity theory seemed less than a century ago. Einstein, as has been widely reported, saw no practical applications of his work, yet lived to see both the nuclear power reactor and the hydrogen bomb. Kahn and Wiener have said that future progress 'may even seem to contradict the laws of physics'*; it has happened before, so why should it not happen again?

That is an extreme case. But between the limits of nothing really new ever being discovered and the whole applecart of physics being overturned lies our reasonable basis for progress – the assumption that *some* new developments will occur, sufficient at least to keep us on the road of growth towards a more equal future over the next critical fifty to 100 years. The vital problems to be tackled are those of ensuring that continuing technological change is applied beneficially, and, in particular, that progress does not continue solely in the rich countries, taking us along the high growth/less equal path to likely destruction.

Society and Change

Although technical change will certainly continue, the way new technology is developed and applied in relation to society as a whole is now quite different from the pattern of a few decades ago. When R&D took place on a small scale and innovation seemed to be produced by cheap and easy methods, there were few questions asked about the nature of research and where it 'ought' to be going. With a few Edisons and Fords around, turning out innovations, society was able to take what it wanted and leave the rest. But now, especially since the Second World War, all that has changed. R&D has become a major part of the national expenditure in the developed world, typically as much as 2

* *The Year 2000*, Macmillan, London, 1967, p. 69. If you want to check out just how far the bizarre extremes of *present* physical theory are removed from our commonsense world view, try my own *White Holes*, Delacorte, New York, 1977.

per cent of the Gross National Product. With huge sums invested in specific projects, the feeling of the investors – either industry or government – is that the public must take the end product whether they want it or not. And, as the thrust of research becomes concentrated on fewer, bigger projects, there is less to 'choose' from anyway.

Big science has become a State activity, with little room left for the individual working outside the mainstream. And, at the same time, nuclear issues in particular have focused attention on the broad range of issues associated with the introduction of new technology, so that there is a growing demand for proper 'management' of research, management which itself adds to the cost of the whole business.

Real costs of producing really new products and processes have been rising across the whole range from new drugs to new kinds of nuclear reactor. In some cases the research costs are so high that not even developed countries can afford them individually, and we have the example, among others, of several European countries joining together to carry out research into nuclear fusion through the JET (Joint European Torus) project. The optimists can, and do, argue that there is still scope for a real breakthrough which could lead to cheaper solutions to some of these problems, and they usually cite the example of the transistor (in relation to the older valve systems) to support their case. But so much of the cost of new technology is attributable to increased concern about, and care for, safety, the environment and other social aspects that it makes more sense to accept that the days of cheap breakthroughs must be, by and large, over. If this pessimism proves wrong, of course, then we will be better equipped than we expect, which is the kind of unexpected outcome we can cope with.

This makes it all the more important that society as a whole should play a part in deciding where the major new investments are going to be made, and that the limited resources for investment in R&D should be used wisely.

Society as a whole is affected by these choices and the resulting change; society as a whole should contribute to discussions on which policy decisions are made. As a first step, we need to identify the critical areas where a correct application of resources could be most beneficial, but a mistake made now could prove disastrous.

The Key Issues

The three main keys to the future have already been identified: food, energy and raw materials. But where should our priorities lie? Which of these issues is of the greatest importance and where within each of them should the main thrust of research be concentrated?

The least of our problems, in these terms, seem to be those associated with the continuing supply of raw materials. In the past, it has proved possible to respond very quickly to shortages of any one material, and there is such a wide range of possible technologies, with ample prospects for substitution, that it really doesn't look as if a major breakdown of the whole materials sector is likely. But there are still urgent requirements for improvements in the system. Can we develop mining techniques which are less demanding in either capital or energy terms (or both)? Can the efficiency of processing be improved? And how best can we make more efficient use of the materials, including by recycling, once they have been mined and refined? One shift in attitudes is clearly indicated, and serves to highlight the overall pattern – we need to look closely at the whole business of obsolescence (planned or otherwise) and he 'life' of products, where, perhaps, a slightly greater initial capital cost can be more than offset by the resulting increase in useful lifetime, whether the product is an electric light bulb, a car or a power station. Even here, however, the right decisions may not be obvious. Longer life per product may mean less production and more unemployment – is that 'good' or 'bad' overall? Some of these implications will be discussed in Chapter Seven.

165

The SPRU team feels that there are many different ways in which future material needs can be met, but that only a broad direction of research into these problem areas will be enough to avoid breakdowns in supply. In agriculture and food production, the problems are sharper and the directions for future efforts to be concentrated more clear-cut – although in the case of both food and raw materials it makes sense to build up stockpiles wherever possible as insurance against disasters such as natural catastrophe (climatic shifts and increased weather variability, say) or war.

Again, in agriculture the underlying theme must be an emphasis on methods which are less capital- and energy-intensive than those of the developed world today. Increased population can often be absorbed usefully by labour-intensive agriculture which produces, if carried out properly, very high yields. To aid this process, the research and development effort should be geared towards bio-logical techniques of pest control, developing food plants that 'fix' nitrogen into the soil (instead of using energy-expensive fertilizers), developing cheap, robust, small-scale tools and equipment, and, no less valid for being so widely touted, techniques for ocean farming and dietary changes.

The food problem is bigger than the raw materials problem, but its solution lies as much in social change as in technical change, as we saw in Chapter Three.

But as the emphasis on reducing energy demands in each of the other key areas has shown, the energy question is the one that must have the highest priority. This is not to say that we should dream of putting all our eggs in one basket and concentrate solely on the energy problems: we must still have raw materials at reasonable cost, and you can't eat electricity. Nevertheless, the possibility of a shortage of energy is the most likely constraint on the whole pattern of growth towards a more equal society that we are committed to. So the concentration of a major part of available R&D effort on the energy problems discussed in Chapter Four is not only justified but necessary.

Very briefly, these important areas for immediate and continuing research are those of conservation (especially in transport and use of waste heat from power stations), more efficient (mechanised) coal mining and techniques for converting coal into gas and 'petrol', storage and transport of natural gas, efforts to locate more reserves of oil and gas, use of solar and wind power wherever possible, use of 'fuel crops', and, as a long-term, slow-developing possibility, thorough testing of nuclear reactor designs as alternatives to the sodium-cooled fast breeder (such as the gas-cooled reactor and thorium breeder). The crucial *mistake* which might be made in the energy sector – itself the most important of our three key sectors – would be to concentrate R&D effort on allegedly 'sophisticated' nuclear reactor systems, such as the sodium-cooled fast breeder or the fusion crock of gold, at the expense of more mundane, but relevant, research into such practicalities as conservation and making best use of the available reserves of fossil fuel.

At least, in the most crucial area of all, we know what *not* to do; the fact remains, however, that there is still a vociferous and powerful lobby trying to push us along this dangerous path!

Catching Up: Technology for the Poor

In terms of social welfare, it is open to argument whether or not technical change in the past few decades has improved the conditions of those living in the poorest nations. Research and development in agriculture, for example, has had an overwhelming emphasis on the economic factors considered important by large-scale farmers in the developed world, with only a small proportion of research on the particular problems of the poor. In this and other areas, including our key area of energy, growing awareness of this problem has produced questions about the 'appropriateness' of Western methods and

Western technology to the problems of the developing world. Put in those terms, the problems of finding ways for the poor to catch up can be easily highlighted: the 'Western' approach depends on lots of capital investment and minimising the labour force, which is the last thing a country that is poor in economic terms but has a large potential workforce should go for.

Although the 'inappropriateness' argument does now get aired, there is still far too little attention paid to such problems in the industrialised nations of the world, and as a result of the lack of discussion the 'answers' to the problems are not yet always apparent in a clearcut form.

But that shouldn't stop us talking about the problems – indeed, it makes it all the more important to get the ball rolling by tossing out some suggestions so that answers can be found by knocking these ideas into better shape, and in the process perhaps uncovering some new and better ideas.

It is in this spirit that we should look at some of the proposals put forward by the SPRU team – not as definitive 'answers' in themselves, but pointers to the right direction to follow, and a basis for the decisions which must be made as we progress along the road. The general problem so far can be linked with the way 'aid' policies in the developed countries are not only separate from policies on trade and investment, but often so different that the two kinds of policy are contradictory. Aid may be used to train people in the developing countries with technical skills, but the use which can be made of those skills within the developing countries depends on just what kind of technology is 'sold' to them by the rich, and on what strings are attached to the deal.

Or look at the example of capital aid – this can be offset in terms of impact on the balance of payments, for example, by establishing a monopoly position which forces the developing country to spend its capital in the donor country. An obvious example is providing a poor country with a cheap railway system, but one which operates with rolling stock and other equipment which can only be replaced and serviced from the country which provided the 'aid'. That is the worst kind of example, since no cunning motives need lie behind the situations which have often arisen simply because the people in government responsible for foreign policies on trade and investment are out of touch with their colleagues responsible for aid, running independent and often incompatible policies – the left hand of government, when it comes to dealing with the developing world, all too often really doesn't know what the right hand is doing.

This is not an easy problem to solve – it needs much more than just new lines of communication between government departments. But the SPRU team feels that efforts to remove the contradictions affecting aid are at the heart of the issue if the developing world is to 'catch up' in a reasonable time, and therefore this must be one of the key political challenges of the next two decades. The USA might take the lead in this situation by extending anti-trust laws to overseas operations, especially in the poorer countries; everyone involved in aid programmes could improve prospects by encouraging the build-up of innovative capability *within* the poor countries, so that they can develop their *own* 'appropriate technology'.

Such policies involve major changes in political attitudes. The countries which can make best use of aid, in terms of shifting the world along the path to equality, are those where the internal policies are geared towards large-scale employment and greater social equality. But these are not necessarily the countries which receive most aid at present, since the existing system helps the friends of the rich without such criteria being used to decide who can best use the aid. Aid policies should also be to *reduce* the costs of using local, as opposed to imported, technology – again a major shift from the present position, and one involving training of local people as well as paying for local research and development and encouraging local production of machinery in the poor countries.

169

So although the problems of growth into a more equal world are not insoluble, and are, as we have seen throughout this book, essentially 'political' as opposed to 'technical', they are still very real. A whole new political attitude and sense of will is needed if the required effort is to be directed towards solving the right problems of the poor countries. The problems of technical change are inseparable from the wider issues of politics and economics, even though the catching-up process can take place within the framework of any of our three basic worldviews. We must look not just at the possibility of providing the necessities for life, but at the quality of life that results in our choice of future worlds.

Before moving on to look in detail at the social implications of the changes we are about to experience, however, it does seem appropriate to look in some detail at the policies appropriate to providing the necessities of life in at least one of our three key sectors. The most important sector, as I have stressed, is certainly that of energy. But the sector in which the most immediate changes can be made, with immediately apparent benefits, is the one that has also been most widely misunderstood throughout the entire history of the 'futures debate'. So I shall choose the food sector as my 'case study' of the policy changes appropriate to a more equal world.

Food for the Poorest

Growth alone is not enough to reduce automatically the poverty which is the root cause of hunger. In many parts of the world, even fairly high rates of economic growth over the past two decades have not directly benefited the poorest, and in some cases the poor have got poorer in absolute terms, as well as relatively, while the rich have got richer. The examples of regions and countries where this has not occurred, however, clearly signpost the right road, the one along which we must travel if high growth is to be the means for producing a more equal world in which the problems of poverty and starvation are first reduced and then eliminated.

When people have enough to eat, they have enough energy to produce a surplus for sale or exchange, or to work in other activities than raising staple crops. So a society in which the poorest are adequately fed needs to grow in other ways, to provide outlets for this productive capacity. Deliberate policies are needed first to reduce existing inequalities, then to ensure that new inequalities do not arise in the new kind of society that emerges with industrialisation. Such policies can be imposed from outside, as other policies have been in the past by physical conquest, the economic muscle power of the rich nations, or by politics and propaganda. Alternatively, the policies can come from within the poor nations themselves, but this will only occur if the people – including the poorest people – have information about what could be achieved and how, about where to obtain advice and equipment, and so on.

Education is clearly important. But, as agricultural economist Keith Griffin has pointed out, the needs of the poorest people can only be met if the basis of wealth is distributed – in other words, if everyone has a share in the land. Usually there is not enough land for each family to have its own viable plot, but each can still have rights in land which is communally owned. The peasant working for himself (instead of for some distant landowner, or simply struggling to reduce indebtedness to a bank) is encouraged to increase productivity and then benefits from improving production directly. Initially, efficient labour intensive methods can increase production of food in many parts of the world, as we saw in Chapter Three. Then, with increasing wealth it is possible to bring in more capital intensive techniques, but while maintaining as the overall objective maximising the yield per hectare rather than maximising the cash return or minimising the size of the labour force.

In the very short term, at an international level, it is still urgently necessary to establish emergency food

171

reserves, perhaps managed by the FAO or some other UN agency, which can be distributed rapidly and effectively to areas where there are real shortages of food. The next step is to use local labour and local resources to increase food productivity in those areas where shortages occur – and for this immediate aim it doesn't matter if the food produced is from animals, even though in principle it is more 'efficient' to eat grain. Semi-desert regions can support some animals where they cannot support crops all the year round, and it is best, in any case, to build on traditional habits and methods. The animals are also useful in other ways, perhaps providing transport and manure for use as fertilizer or fuel as well as producing meat and milk. Let's face it – the way things are today, even if the people of the rich countries simply stopped eating meat, no huge surplus of grain would be released to feed the hungry. The grain which used to be fed to beef cattle simply wouldn't be grown at all, because the poor wouldn't be able to pay for it on the world markets.

Where the poor have become less poor, they are able to change world trading patterns and benefit from the surplus capacity of the rich. In the USSR and Japan, for example, very little meat was eaten until quite recently – indeed, in many cases very little of anything was eaten. Once the USSR exported wheat while the poor ate beet, cabbage and potatoes; now, their meat consumption is approaching the European levels of the 1950s, a situation only made possible by importing grain from the USA to feed the animals which provide the meat. Earl Butz, US Secretary for Agriculture, claimed in 1972 that this new market had 'saved America's mid-West' from depression – rather than the increased wealth of the citizens of the USSR putting a strain on world food production, it has helped to keep many farmers in business! And the fact that equally dramatic changes in eating habits have taken place in Japan highlights the important point that a better world can be achieved, with a reduction in poverty, within a very broad range of politically-flavoured scenarios.

172

What are the farming practices which could be used in developing countries immediately, at no cost and using only advice and information? Plant cover should be maintained throughout the year by sowing the same crop at different times and by using different species: bare soil, when it occurs, should be covered by mulches of leaves, brushwood or even plastic sheeting to trap moisture and protect young plants; and all available sources of plant nutrient, ranging from sewage to ash and compost, should be utilised. Such actions sound too simple and 'obvious' to make much difference – yet all too often in the developing world agricultural producers abandon such good habits and simple, straightforward methods in attempting to make the leap to single crop larger fields with mechanical aids, with the small farmers displaced and financial profit, not maximum yield, the over-riding motive.

Slightly more initial effort would allow further techniques to be used to increase productivity: identifying *specific* nutrient requirements in the soil of a region and using fertilizers tailored to meet them, as has been done very successfully in Ireland; similarly identifying irrigation needs; finding outlets for surpluses and providing cash credits where necessary; and, of course, land reform to give the farmers a direct personal interest in increasing productivity.

In some regions, terracing of hilly land – potentially suitable for good production – requires more labour than is available, and there is a need for small, rugged machinery to do the work – *not* for huge, expensive super-tractors that cost a fortune and only work in big, flat fields. Here, as with the many similar examples, is an obvious way to develop emerging industrial capacity and the local innovation and research that are a necessary stimulus to the healthy growth of the poor nations.

The underlying problem, however, remains. In order to be fed, people need to be able to exert economic and political leverage; but in order to achieve economic and political power they need to be adequately fed and productive. The first step to break this vicious circle must, it seems, be a humanitarian one. The rich nations must help the poor. Long-term strategies have already been debated and are included in the declared aims of the UN and its agencies such as the FAO and WHO. Can the governments of the rich nations be persuaded to move the UN and its agencies along the paths already mapped out, providing them with the resources to do the job? Somewhere, someone must make the first political step; if they do, the SPRU team suggests that the already declared UN aims could be carried out roughly as follows to reduce inequalities and starvation in the world:

1. UN agencies authorised by General Assembly to co-ordinate food policies so that wealthy and powerful nations, or blocs, cannot dominate others.
2. Agriculture planned nationally and regionally to use existing knowledge to conserve soil and increase fertility and productivity per hectare wherever possible.
3. Research into potential food plants to be encouraged by the UN and supported by governments, following the successful similar research into protein sources.
4. Investigation of plants that can fix nitrogen from the air into the soil.
5. Increased efforts to make use of local knowledge about food resources, and development of these as soon as possible, instead of sweeping away traditional methods and putting in new crops regardless of circumstances.
6. Encouragement of ranching of indigenous wild animals – even in the developed countries there is potential here, including experiments now going on with deer farming in the UK.

Many of these developments could take place alongside one another, and all should be seen as being set in the context of land reform, accepting the need for changes in society that are implied. This example from the food sector emphasises again the conclusions reached so far in this book regarding the possibility of achieving a desirable future world. We know what needs to be done, and we know that it *can* be done. From here on the problems are essentially political, and the debate about the future becomes inescapably a political debate. Clearly, this is the time to look again at the different 'worldviews' which we came across earlier in the book, but armed now with a deeper understanding of the nature of the physical problems that must be tackled within any chosen worldview.

Worldviews Revisited

The three worldviews that we met in the early part of this book represent three political standpoints: conservative, reformist and radical. Each of these represents a different approach to understanding how the world works and what should be done to keep it working, or to change the way it works. The background to these standpoints has been the stuff of many books, and it would be inappropriate even to try to encapsulate the economic arguments here, although a comprehensive discussion has been provided by the SPRU team in *World Futures: The Great Debate*. But a few points should, perhaps, be highlighted.

It is particularly important to realise that as circumstances change then economic views, even within each school of thought, themselves change and evolve. A particular specialist writing at a particular time might produce a piece of work which could undeniably fit into one of the categories 'neoclassical', 'Keynesian' or 'Marxist'; but the same specialist might produce quite different pieces of work at other times, and the boundaries between those three categories might themselves shift as time goes by. The 'Malthusian' debate provides one example; what Malthus had to say meant one thing within the economic, political and technological environment of the early nineteenth century, but has quite

175

different implications when the same questions are viewed from our standpoint in the last quarter of the twentieth century.

And, of course, even within each broad category at any one time there are differences – often sharp differences – between the views of different people. So we have to pick out the continuities within each broad tradition. Does the worldview we are looking at concentrate on the roles of individuals, classes, or nations, for example? It is the continuity of a theme within each worldview that determines how proponents of that view decide what is and is not possible or plausible developments in the world. This has direct physical bearing on the problems discussed in earlier chapters. The extent of proven reserves of material resources in the year 2000 will depend on how hard we look for resources, which in turn depends on the political attitudes and worldviews of those who have power to encourage, or to discourage exploration. Estimates of the ability of society to overcome the kind of difficulties that the world will face over the next 100 years affect the way in which society actually tackles those problems, and colour the overall reaction to potentially damaging surprises such as natural disasters. A comparison of worldviews lies at the heart of the question of *how* society will respond to the problems coming up in the years ahead, and therefore at the heart of any study of alternative futures.

The key issues here concern the relation between the three worldviews and the prospects for growth and equality which are central to the SPRU view of a desirable future world. Again, other issues and a full range of scenarios are covered at length in the SPRU work, but here I shall deal only with the issues central to the present book. Both kinds of inegalitarian future (high or low growth, less equal) can be dismissed, as before, on the basis that they lead to increased tensions and heighten the prospects of conflict – and that, with three worldviews to choose from, rules out no less than six options at the start! The three low growth/more equal scenarios deserve perhaps passing mention, because there are superficial attractions to this Robin Hood kind of approach, and it has received a great deal of attention recently.

If the rich countries of the world were to grow only slowly, if at all, in economic terms from now on, and if the developing countries grew only until they had 'caught up' to this standard, we have the family of low growth/ more equal future worlds. But the proponents of such an alternative have different reasons for putting it forward, and they have different ideas about the right level for growth to stop at.

For conservatives, such a future would come about not as a 'best possible' option but through a failure, in some sense, of the present system. War, political conflict or climatic change might hold back the growth of the rich north while allowing the poor south to develop, or, more likely, economic stagnation in the north would result from changing cultural attitudes, apathy towards work as a 'good thing', trade union power, increased bureaucratisation of the state, and so on. In many ways this is the future foreseen by the gloommongers – things falling apart basically through bad management and creeping socialism.

Reformists, on the other hand, could see a low growth/ more equal future arising as a result of a deliberate redistribution of world wealth, a belt-tightening in the north coupled with a fairer share of the cake we've got already. This would be the best possible world for some theorists, with the emphasis on survival as a better alternative than destruction through conflict or through a ruined environment.

And the radicals would see things differently still. Their low growth/more equal world might come about through the imminent economic collapse of the rich nations, accompanied by a mounting challenge from the developing world. Now, the situation is seen as neither desirable in itself nor as the ultimate goal, but as a prelude to better things, to the high growth/more equal world which, in their view, could follow after the collapse of capitalism as a socialist world emerged from the remains.

None of these options stands up as a desirable alternative to a high growth/more equal future world, provided that can be achieved without conflict and ecological crisis. But the examples help to highlight the differences between the three worldviews, setting the scene for the three more detailed scenarios that offer the best hope for a stable, egalitarian world.

Three High Growth/More Equal Scenarios

All three versions of the high growth egalitarian (HE) future world depend on a high level of world trade, the means by which the bigger cake is shared more equally, but they differ in their assessments of how this situation should come about. Conservatives simply see a need to reduce or remove trade barriers and let the forces of the market place work as they will to provide the benefits of production for all; reformists suggest that the forces of the market on their own will not necessarily act to ensure that the poor, as well as the rich, get richer, and see a need to restructure the world economy to produce a more egalitarian distribution of wealth; and the radicals argue that such planning could never succeed against the entrenched interests of commercialism, so that only by revolution can the world system be changed sufficiently to ensure a fair distribution of available resources and the benefits of industry. These three lines of argument are spelled out in Figure 26, and we can put flesh on that outline along the following lines, taking each scenario in turn.

Conservative: The touchstone of this worldview is that competition in a free market ensures maximisation of output. In a future world where this principle has become widely accepted and put into practice by governments, there would be a shift away from policies aimed at bolstering declining industries in the face of competition and no further attempts to control exchange rates to bolster the national prestige and self-image. Multinational corpor-

178

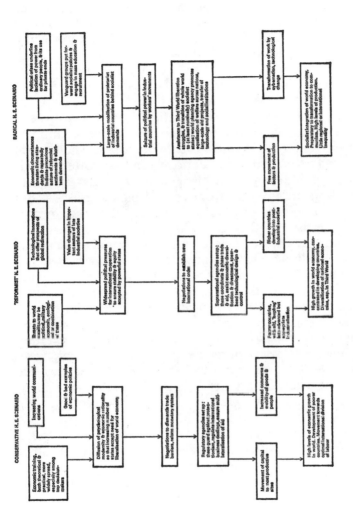

CONSERVATIVE H. E. SCENARIO

Economic training, both theoretical & practical, more widely spread, especially among top decision-makers

Increasing world communications

Good & bad examples of economic policies

Diffusion of psychological modernity & economic rationality so that increasing number of states accept need for liberalisation of world economy

Negotiations to dismantle trade barriers, reform monetary system

Regulatory agencies set up: these guard against protectionism, regulate international business dealings, ensure multi-lateralisation of aid

Increased commerce & mobility of goods & people

Movement of capital to most productive sites

High levels of economic growth in world. Development of poorer countries. Movement towards optimal international division of labour

'REFORMIST' H. E. SCENARIO

Threats to world stability—may be political, military, economic, ecological or combination of these

Technological innovations that offer prospects of global redirection

Value changes in important sectors of late industrial societies

Widespread political pressures for international cooperation to ensure stability & equity accepted by powerful states

Negotiations to establish new international order

Supranational agencies set up: these coordinate & phase trade & aid, assist economic diversification & investment, spearhead technological design & control

Poorer countries, with aid, 'leapfrog' backward but accelerate industrialisation

Richer countries develop into post-industrial societies

High growth in world economy, concentrated in developing countries, but throughout. New economies, esp. in Third World

RADICAL H. E. SCENARIO

Political crises underline isolation of power from ordinary people, & its use for private ends

Economic circumstances threaten living standards & repeatedly indicate precarious nature of reformist ambitions & short-term demands

Vanguard groups put forward socialist policies & engage in mass education & recruitment

Large-scale mobilisation of proletariat of industrial countries behind socialist demands

Seizure of political power in industrial countries by workers' movements

Assistance to Third World liberation struggles, & transition of whole world to (at least nominally) socialist states: world planning agency promotes egalitarisation of welfare across nations, transfer of technology and social institutions

Free movement of factors & production

Transformation of work by automation, technological change

Socialist integration of world economy. Preparatory to transformation to communism. High levels of production. Little regional or international inequality

Figure 26 : High growth, egalitarian scenarios

ations would be seen as important contributors to development through their investments in poor countries, with the poorest people benefiting automatically through resulting economic growth. By and large, this is the picture that Kahn's team sees as the best possible future world, and describes in *The Next 200 Years*.

The changes implied would be large, and could take place only slowly, over several decades, even with a deliberate effort by governments to break down trade barriers and reform the international monetary system. Taken to its logical extreme, the conservative solution to problems of inequality must eventually see the whole world returning to a free market economy, and supporters of this view point, for example, to the prospects of East European countries shifting away from central planning as a result of unrest about living standards expressed by their people as they realise that they are falling behind the West.

Reformist: The key difference in the reformist scenario is that the multinational corporations are no longer seen as a necessarily benign influence on development, so that economic co-operation between countries is needed, on a planned basis, in order to produce the HE world. Multilateral agencies, rather than multinational companies, would be the means to regulate investment and aid, and to guide trading developments, perhaps with an emphasis, initially at least, on regions of the world rather than the globe as a whole. But reformists recognise that the required changes cannot take place without some accompanying conflict.

The present rich countries would be steered into a 'post-industrial' state with emphasis on the quality of life rather than the quantity of material possessions, and the developing countries, learning from the mistakes made elsewhere, would avoid many of the most unpleasant features of industrial growth that the rich have already experienced. In the very long term, with increasing co-ordination of social and economic planning of different countries, such a future world would establish some form of world government, whereas in the conservative HE world that would hardly be necessary or appropriate as the functions of government, if anything, withered away as progress was left in the hands of trading companies and to the forces of the market place.

Radical: As the name suggests, the radical view involves a drastic change in the way the world is organised, sweeping changes not just in the capitalist countries but also in many of the countries that describe themselves as 'socialist' today. Open class conflict as a prelude to the establishment of a new socialist order in the industrial companies would coincide with, and encourage, liberation movements in the poor South, a period of turbulence that can only be described as a 'world revolution'. Only then, with power in the hands of the labouring masses, could development be planned on a basis of human needs rather than according to the interests of big business.

After the revolution, however, the situation might in many respects echo the world the reformists see coming about through evolution, not revolution. World planning agencies would still be required to co-ordinate growth into a more egalitarian society, but now with large firms nationalised, and land reform, including collectivisation, at the heart of development strategy in the non-industrialised countries. Once again, the developing world could learn from the mistakes of the present rich, with the implication that the now-socialist rich nations would be willing to provide massive aid to the poor, while at the same time in the long-term interest the socialist poor would accept restraint on consumption until production capacity was built up, rather in the pattern of China in recent decades. Now, the key difference from either of the other scenarios is that growth would take place to meet social needs rather than for profit, with the role of money itself eroded as increasingly basic needs were provided free of charge.

These three scenarios, as indeed the nine others outlined by the SPRU team, are oversimplifications and cannot be

181

regarded as realistic possibilities. Some mixture of the visions from the oversimplified worldviews is inevitable, as also is some mixture of high and low growth, equality and inequality, as different regions of the world pursue different courses in the immediate future. But the variety of scenarios helps to point up the significant factors which are going to affect future trends, a framework within which policy decisions can be put in perspective so that the implications become clear. The fact that some version of an HE future world can be envisaged in *any* worldview is more important than the detailed way in which a literal interpretation of any one worldview maps out the path ahead.

In their own work, the SPRU team has neither estimated the probability of any one of the twelve possible scenarios being closest to reality, nor indicated a preference for one kind of future world as a desirable option. In encouraging me to write the present book, the team has shown its colours to the extent of approving my choice of the high growth/more equal options as a focus for attention, the range of possible future worlds which is clearly the most desirable of the four possibilities covered by growth and equality alternatives. Within that range of options, it seems to me that from past experience there is no evidence to support the view that market forces alone can produce more equality, as the work by Griffin on food has shown. There are parts of the world today where economic growth has made individual countries more wealthy, while the poorest inhabitants of those countries are now worse off *in real terms*, as well as relatively, than they were before. So it is my personal opinion, which should not in any sense be taken as the consensus of the SPRU team, that of the three scenarios outlined in this section one, the complete *laissez-faire* market place economy of the hard-line conservatives, is not likely to produce a high growth/ more equal world in practice.

Introducing any element of control of the market forces makes the scenario, strictly speaking, a variation on the

182

reformist theme, so that in this sense our choice is restricted to the reformist or radical paths. One last question remains: what would it be like to live in one of these future worlds?

Chapter Seven

Living in the Future World

Even though there is no attempt in the SPRU work to describe what 'the' future will be like, it is still possible to look at some of the main implications for lifestyles in the whole range of high growth/more equal future worlds described in the previous chapter. The basic theme of any such general assessment of what it might be like to live in the future world must be the quality of life, since growth and possession of the fruits of technology do not alone produce 'the good life'. But the three different windows through which we may look to obtain our three different views of the possibilities for the future do not look out upon the same scenes. At best, they look at different aspects of some broader scene, much of which remains concealed from view.

Depending on which worldview you subscribe to, your assessment of what makes the good life good, may see equality, or liberty, or preservation of the natural environment high on the list of priorities; some idea of the differences – and in some cases similiarities – between the worldviews is provided by Figure 27, where seventeen issues related to quality of life are viewed through the different political windows. The list is not meant to be comprehensive, but provides ample background for a detailed look at two of the most important, and complex, issues of quality of life, those of liberty and equality. In spite of the rallying cry of the French revolution, it seems that in some political frameworks these two desirable bases for the good life may be seen as working against one another, with a very crude caricature of the range of opinion setting liberty at the heart of conservative philosophy (with little or no thought for equality) and equality at the base of the radical world, even at the expense of liberty, with reformists somewhere between, offering restricted liberty and if not equality a reduction of inequalities.

Such a caricature may be useful, like a mnemonic, in providing a basic outline for easy remembrance. But it certainly does not tell the whole story. So perhaps we do need to look in a little more detail at the political background before moving on to speculate about the way of life in the future world – the implications for work, housing, health care and so on.

The Political Framework

The conservative high growth future (with more equality almost as an incidental bonus) is best described by Kahn's team, which sees the crucial shift into less rapid growth in both world population and 'Gross World Product', the demographic transition around the bend of the 'S' shaped curve, taking place by the mid 1980s. They may be right – there are already signs that the growth of world population is indeed slowing and that the spectre of continuing *exponential* growth is laid. But even Kahn sees problems in the short term, not least because of the possibility of a premature halt to growth as a result of the fears raised recently by what he calls 'the prophets of peril':

> If enough people were really convinced that growth should be halted, and if they acted on that conviction, then billions of others might be deprived of any realistic hope of gaining the opportunities now enjoyed by the more fortunate . . . eventually – when the post-industrial economy has arrived – much of the industrial imperative and its appurtenances will erode or expire;

185

Figure 27: Seventeen issues, related to the quality of life in a high growth egalitarian future world, looked at from the perspective of our three chosen world-views. In many cases, the implications for the developed countries (LDCs) are assessed in parallel. Based on data from *World Futures: The Great Debate*

	Conservative	Reformist	Radical
1. Geopolitical Issues	International agreements gradually dismantle trade barriers, multilateralise aid, reform monetary systems, etc. As national boundaries diminish in importance, there are shifts towards an optimal international division of labour, with much trade and mobility of capital and people. Culture and tastes become increasingly global. Multinational corporations transmit enterprise and skills around the world.	Combination of domestic political pressure in DCs, global political or economic threats, and opportunities offered by new technology result in international agreements to stabilise resource prices, and co-ordinate expanded trade and aid interchanges. An economically diversified world would emerge reflecting local diversity as well as specialisation.	Period of struggle and turmoil, culminating in metropolitan and world revolution. Likely to follow on major economic crises of capitalism, but only with adequate preparation of socialist movements and might be sparked off by 'political' events such as new colonial confrontations undertaken amid recession. Socialist states formed in DCs, move towards internationalism: eventually set up world planning agency pooling resources and knowledge and transferring skills and technologies.
2. Political Development and Change	The global trend is towards establishment of liberal pluralist stages. In many LDCs the emphasis of regimes may be puritanistic, with religious overtones. In DCs there would be much more tolerance of differing life-styles. State regulates Unions and other particularistic	DCs: political power balance affected by pressures from the young, from professionals and technical workers, etc. Much delegation of responsibility to local authorities. State able to attend to national and global growth and permit redistribution of wealth. LDCs: progressive	DCs workers' states, with national goals determined by delegates rom workers' and community councils. Strong central governing bodies but under public control and scrutiny and with much regional autonomy. LDCs socialist states, possibly with some remnants of state capitalism which are being phased out; development

	interest groups, ensures welfare of infirm and aged, and prevents discrimination against individuals or products based on race, national origin, etc.	forces and sectors of society challenge traditional elites, development proceeds with more attention to local accumulation and less cost to the poorer sections of society.	towards communism underway.
3. Social Stratification, Status and Power	DCs: wage rates would be similar in different countries, entrepreneurs and business leaders highly respected. Power diffused among many heterogeneous social groups. Occupational stratification would be marked but only one of many bases for social differentiation. LDCs: as above except: more income inequality, traders and small businessmen highly regarded, pluralism still somewhat rudimentary although traditional elites waning.	Almost everywhere technical experts, associated diplomats and statesmen, are highly regarded. Otherwise, high status in DCs to local organisers and businessmen, in LDCs to state officials. Professional and social service work favoured careers. However, in the salience of occupational status may be reduced, since needs for status are being replaced by more pride in craft work, concern with community.	DCs and LDCs: labour much praised, but mental work still higher status, and power may be linked to service to the Party. Continuous attempts to reduce inequalities between regions and labour, but slow erosion of established cultural attitudes.
4. Distribution and Equality of Material Living Standards and Wealth	While some Southern countries are still relatively poor, and national levels of wealth differ quite substantially, there is globally an unprecedented affluence. Large wage differentials in all countries; pockets of poverty in some.	DCs have established, and LDCs are moving towards, a basically meritocratic society with moderate income differentials, and probably graded increases in wealth through the career. Minimum wages, and the free or subsidised provision of many of the basics of life, would be likely ways of protecting the poor.	Equalisation of living standards proceeding across countries in terms of occupations. Attempts to remove manual/non-manual and town/country distinctions. Abolition of money on the agenda as goods are produced on basis of social need.

	Conservative	Reformist	Radical
5. Social Values and Perceptions	The world is moving towards a society of leisure, with discontents stemming more from impersonal conflicts than from a sense of social injustice. Some 'alienation' in DCs expressed in cultism, sensation-seekings and the like; some protest about change in the LDCs expressed as trades unionism, revival of traditional dress and customs, etc.	Sense of identity established by exaggerating cultural styles, life-styles, etc., but also ethos of restraint dominant. Attitudes liberal (or liberalising) towards sexual behaviour, and non-harmful deviance, and collective responsibility taken for emotional 'casualties' by provision of counselling services, social workers, etc. Widespread belief in legitimacy and responsiveness of 'technocratic' government. The above account is more representative of reality in DCs than LDCs, but the latter are tending in this direction.	Process of building socialist consciousness underway; early fervour may face challenges from pockets of discontent, from confrontation of ideals with 'practical' short-term conflicts. Widespread sense of personal potency and of community power in determining collective interests and action.
6. Action, Order and Conflict	Many voluntary organisations, much interest in leisure and hobbies. Fashions important. LDC culture mirrors that of DCs but slightly less developed. Little crime, due to general affluence and efficient security.	Emergence of diverse lifestyles, much energy expended in 'pursuit of identity' and in finding variety. Experiments with different life-styles and ways of regulating communal organisations. DCs: Participative society, much involvement in local community management, in aesthetic and environmental design, much use of advanced communications systems.	Mass participation in political activity; networks of organisations establishing and linking local, regional and world production, development goals, and social affairs. Much concern with self-education, both through experience of different communities, and through more conventional means of instruction.
7. Social Services and Welfare	DCs: Facilities highly developed, largely under private initiative and	DCs: Health and welfare services fairly egalitarian and collectivised,	Essentially similar forms being developed in already industrialised

incentives of profit as much as of charity; state aid restricted largely to aged and infirm. LDCs: lacking the advanced technology of service delivery used in DCs but able to employ chemotherapy, contraception, etc. Fairly well developed welfare services, especially in industrialised areas.

with a focus on preventive approaches: associated with educational campaigns, environmental protection, minimum wage legislation, etc. LDCs: Much development of paramedical services, state protection of workers' health, collective provisions for aged.

and in developing countries, with many basic goods and services provided free. Stringent health standards at work, preventive medicine, provision for care of elderly and continuing education.

8. Food and Nutrition

In the developed world – a high and varied consumption with imported styles of cooking, and many convenience foods. In LDCs a robust and adequate diet with fish and meat. Because of trade patterns: a tendency to global homogeneity with differences in eating following social status divisions rather than national or regional lines. Possibly some areas still malnourished; probably much over-eating in affluent areas.

The rich industrial countries would have plenty of food available, but the emphasis would not be on high consumption or convenience as much as on variety and the appreciation of good cooking. Some developments of novel farming methods in arid regions, of local meat animals and modifications of local crops. Revival of traditional diets, and cross-fertilisation of cookery across different cultures, possible.

Attention paid in first instance towards production and distribution of food to meet regional standards, with orientation towards upwards convergence of standards. Probably this would imply reinforcement of many national and regional dietary styles, but these would also be affected by the development of farming technologies in previously unproductive regions.

9. Health

In DCs we might expect heavy emphasis on curative medicine, with fast technical advance in prosthesis and the cure of disease. Preventative medicine (in addition to basic sanitation etc.) might well be minimal, but functional exhortations to exercise might be

In this scenario both preventative and curative approaches to health and medicine would be likely. High standards would be set in order to secure protection from harmful effects of foods and drugs; and international FDA might be set up. Concern for individual

Emphasis on preventative medicine and on technological (bionic) means of coping with accident injuries, etc. High levels of protection from harmful substances might be imposed e.g. severe restrictions on smoking; also possibly compulsory medication for some physical

	Conservative	Reformist	Radical
	common in a 'leisure society', thus economical exercise technologies – isometrics, home exercise machines – might be prevalent. In LDCs routine welfare benefits, sanitation, contraception, would be introduced. Probably advanced curative technologies would be for the rich. Massive health campaigns to eradicate particular diseases would be likely.	liberty might, however, limit the effectiveness of such developments. In LDCs gradual development to DC standards via paramedical-based systems is likely.	ma adies. Welfare of community would probably dominate over individual liberty here.
10. *Work*	Automation is being or has been introduced on a grand scale; many forms of manual labour are archaic in DCs where employment may even be seen as a privilege for the highly educated. In LDCs work is likely to be more labour-intensive, particularly in primary sectors. Much work would take the form of information-processing or service employment, unless these activities too were the focus of technological innovation.	In industrial countries the most wearing and boring tasks would probably be automated, but the emphasis is most likely to be on job enrichment. There would be attempts to integrate work and leisure, making the former more a creative activity with group technology, industrial democracy, flexitime, etc. In LDCs technologies would be less sophisticated and work might typically be carried out at factories that resemble an agglomeration of craft workshops. Industry would here be fairly labour-intensive.	Considerable emphasis on automation as a means of reducing the differences between manual and mental work; technologies chosen in order to maximise worker control and choice. A fundamental change in the nature of work, away from its alienated form as wage-labour wou'd be underway; but objective differences would probably persist between different countries, regions and types of work in the extent to which this process has begun.
11. *Leisure*	In DCs people with free time (which will be practically	In DCs and LDCs alike we might expect serious attempts to	In DCs leisure activities largely educational, linked to community

everyone) will find a 'cultural delicatessen' of crafts, hobbies, obscure areas of knowledge, available in heavily packaged and predigested form. Tourism would be an important activity, and a justification for maintenance of the form of indigenous cultures, whose substance has long ago perished.

In LDCs because of 'global village' media effects, fashions mirror DCs; 'crazes' for sports and popular entertainments, particularly film or pop stars, might be frequent.

reanimate traditional arts and local cultural forms. This would probably go hand in hand with cultural eclecticism, recognising equivalent validity of alternative cultures. Much international apolitical (Olympic Games) sporting and cultural contact. Deeper contact with non-Western culture might lead to an inversion of the cultural domination of the mass media, a possible world-wide predominance of LDC philosophies and art-forms. Probably major growth in coverage and scope of mass media, with innovative and 'personalised' communications receivers.

needs, or oriented towards improving mind and body (chess, gymnastics), functional values and realism set above abstract aestheticism. Social analysis might become an artform. LDCs similar to above, more intermingling of leisure with education and work of development. In both cases, the distinction between work and leisure would be fading.

12. *Education*

In DCs education of children is in part education for leisure, in part utilitarian and functionally oriented, probably making use of programmed computer learning techniques as a supplement to conventional classroom technology. Adults might well be frequently retrained according to demands of technological change. In LDCs education is largely based on western systems, using imported packages of materials, satellite beaming of educational TV, and western-trained experts in colleges and universities.

In DCs the goal of education might be the creation of generalist, liberal intellectuals, creative artists and craftsmen. Intellectual pursuits and intellectuals are highly valued. Education lifelong and continuing; some training in social skills provided. In LDCs the orientation would probably be more functionally oriented towards engineering, medicine, entrepreneurial skills.

Education is oriented towards training of basic skills, but with more overt emphasis on personal development; this aims at producing socialist consciousness. High value is placed on productivity and labour, practical relevance; the goal would be to assist people to be part of a collectivity rather than to train them into being rulers and ruled. LDCs might follow the Chinese example, with experience of labour (e.g. actually *working* on the land) regarded as part of a rounded education.

	Conservative	Reformist	Radical
13. Shelter	For the wealthier groups in DCs housing might resemble a high technology UK 'stockbroker belt'/ US 'Westchester County'. The majority of people might live as conventional nuclear families in highly automated, self-contained, inward-turning households, mostly in dense urban agglomerations. Mass produced housing estates would be likely in LDCs with high density flats replacing squatments, and housing standards improving with economy.	In DCs well planned 'garden cities' (or suburbs) designed to maximise privacy, though concern for environment might lead to higher densities of living than at present. Family structures would probably remain basically nuclear, though deviance from this (communal living, etc.) would be tolerated. LDCs would use local materials and traditional building systems to upgrade and finally replace shanty towns and to build new urban centres.	In DCs high density, high amenity, multi-occupancy apartment blocks set in cultivated parkland might be the norm. This would imply high local density of living, with workplaces in close proximity, but ready access to country and parkland. In LDCs there might be an emphasis on communal development, with extended families providing social welfare/security. In this case, housing would be substantial but with limited facilities and extensive communal facilities (e.g. cooking). However, it is possible that socialist aid on a massive scale might allow for the development of new cities along the lines of those described above for the DCs.
14. Transport	In DCs personal transport would probably be mainly privately owned and operated, though with high levels of state provision of roads and perhaps systems to aid automatic guidance. If public transport were to be developed and modernised, this would follow the dictates of profit and be oriented towards the needs of the more affluent: we might find inter-city rapid transit, PRT,	In DCs there might be an emphasis on high technology public transport – super jumbo jets, moving pavements, widespread PRT: dial-a-ride, subsidised shoppers' buses etc. for poorer areas. Restriction on private access to cities – hence 'park and ride' systems – would be likely. Advanced telecommunications probable. In LDCs similarly emphasis on	Transport as a service is provided free or very cheaply, especially in cities. Private cars might be reserved exclusively for out-of-town, holiday leisure use. Because of planning of urban forms, many journeys, particularly those to and from work, achieved by walking. In LDCs we might expect a heavy emphasis on goods transport as instrumental to growth, with possibly some local environmental damage (e.g. that due

self-drive electric cars. Conventional public transport would be rudimentary and concentrated around low income areas in the richest countries. LDCs would be moving along the same axis, substituting private for public. In all countries and internationally, rapid efficient goods transport would be developed to facilitate trade.	public transport bus services, jitneys. Local authorities would regulate privately operated trams, e.g. enforcing car sharing. Public expenditure on traditional technology, e.g. railways, would be likely.	to heavy lorries) for the short-term traded-off against long-term development needs.
15. Energy Issues High levels of consumption in DCs much for personal transport. Energy intensive agriculture. Oil exhausted rapidly, thermal nuclear insufficient, breeders required. Conservation only if commercially necessary. Potential for ecological problems. Not much attention to 'alternative' energy sources. Similar patterns in LDC consumption geared to industrialisation rather than agriculture initially.	High consumption, but some attempt at energy conservation. Long-term planning of energy supply including attempts to 'stretch' the lifetime of oil reserves. Nuclear important (possibly including breeders) but encouragement for solar, hydro, wind power etc. Lower heating and transport demands from 1. Smaller cars, insulation of dwellings, compact cities, and less artificial fertilizer in agriculture. More attention to rural problems than in conservative world requirement for smaller energy installations.	High consumption, oil depleted as in reformist world Rapid increase in consumption with attention to local sources. Attempts to plan as in reformist world but less attention to conservation and environment initially.
16. Materials and Resources Issues High levels of consumption of, and trade in many materials; much increased mining, especially	Increased recycling of some materials, stimulated partly by environmental concern, partly by	High levels of resource consumption and extraction; recycling developed for reasons both of economy and of

in the Third World (unless rising energy costs make recycling economically attractive, or new technologies allow for the widespread conversion of currently unused, but readily accessible and widely distributed, materials into useful forms such as ceramics). High requirements for energy, relatively high degree of environmental degradation, in mining.

increases in materials prices (which are a possible form of indirect aid to LDCs). Attempts to reconcile demands for increased energy with the environmental impacts of many energy technologies, e.g. by improving the efficiency of energy use in transport and heating, development of renewable power sources.

environmental protection. Materials consumption levels rapidly increase in LDCs: much reliance put on developing local materials and industrial bases.

17. Environment

Environmental standards vary with local geography and level of economic development, protection is a response to unwanted side effects. Commercialised beauty spots landscaped high income estates. 'Dustbin' areas from mining cleaned up if necessary. Wildlife and oceans selectively protected for tourism and fish farming.

Pollution-emitting industry located in regions where lower standards are acceptable. Game reservations for tourists.

Attempts to set up world-wide standards, with anticipatory evaluation in a comprehensive fashion. Attempts to preserve and restore wilderness and historic cities on a wide scale. Fair variety in appearance of artificial environment (e.g. agricultural landscape). International legislation on exploitation of oceans etc. Perhaps attempt to restock oceans.

In LDCs attempts to apply international standards and to develop non-polluting appropriate techniques. Spoilation arising from mining world probably be usually restored as soon as possible.

In many respects this scenario might be a mixture of the conservative and reformist views. In DCs standards are applied and 'upgrading' of landscape to improve on nature. Parkland for recreation and wildlife parks and zoos.

In LDCs standards are more selective, and refurbishing less priority. 'Clean up' campaigns conducted periodically and perhaps mobilisation of local environmental corps.

but to weaken it prematurely, before it has run its natural course, would be to impose unnecessary trauma and suffering and make even more difficult the full exploitation of the many opportunities now available. (*The Next 200 Years*, Associated Business Programmes edition, p. 211.)

More about the post-industrial economy later; but Kahn's is the classic conservative view that welfare can best be increased by producing maximum output from economic systems, with those who make the biggest contribution to increased productivity receiving the biggest rewards. 'Liberty' is seen here as freedom of the individual worker, for example, from the tyranny of rampant trade unionism, and 'equality' is the equality of all people in law and their equal responsibility to behave in a socially acceptable way with no rocking of the boat. Life in *this* future world is best seen by looking at domestic policies, where although conservatives would agree that such services as law enforcement are best provided by the State, there should be, compared with the present situation, a decreasing involvement of the State in the running of society and the economy, at least in the developed countries.

For the individual, the conservative emphasis on liberty would be seen most clearly in the restoration or maintenance of choice in areas such as education and health where many governments now extend their influence. The increased efficiency and higher standards which they expect to result from the operation of the free market system are seen as just as relevant to heart surgery, say, as to automobile production. And the inverse of these arguments also holds true in the conservative worldview – present ills, including economic stagnation, the inefficiency of health and education services and failure to provide adequate housing are seen as the result of too much bureaucratic State intervention which stifles the potential of the free market system.

By and large, reformists accept a great deal of the underlying arguments of the conservative view, but they do not agree that these should be applied without modification in the real world. The social theory may work in isolation as an ideal case, but since the world is not ideal changes have to be made to make it workable in practice.*

So in the reformists' world there is always a balance to be struck between freedom and equality. Because the economics of the real world do deviate from the perfect running of the idealised conservative theory, intervention is often necessary to keep things moving in the right direction. An unguided social system is no more likely to find the road to the good life on its own, in their view, than an unguided car will find the way home for you. The guidance they advocate covers areas such as health and education, seen by them as so big that they are best managed by the State, and may extend to regulating the behaviour of both business management and the trades unions by some form of incomes policy. Neither trades unions nor big corporations are seen as inherently evil, but both are to be constrained so that while inequality remains in the world the range of inequality is reduced, with even the poorest having certain basic essentials for a reasonably good life, and the further fruits of a smooth running and productive economy going to a meritocracy made up of those who work harder, who have extra responsibility, keep particularly essential services running, and so on.

In global terms, one of the key arguments of the reformist school is that completely free trade *now* would benefit the richer, stronger countries and be disadvantageous to the developing world. Deliberate effort is needed if the poorest nations are to achieve even the basic essentials of the good life, and such an effort can only come from governments or international organisations

* This is rather like the situation in physics, where it is often useful to start out with an idealised picture, such as that of atoms as 'hard billiard balls' bouncing off one another, which gives a clear picture that makes the basic theory easy to understand. Then these idealised basics are modified to take account of how real atoms behave in the real world.

196

which are ultimately answerable to governments. But it's worth pointing out that while many politicians (especially those in developing countries) prominently support reformist views on world trade, all too often the same people are very reluctant to follow a reformist road with their own internal domestic policies! It is this kind of difficulty in getting reformist control of the conservative economic machine to work that encourages our third group, the radicals, to go the whole hog in proposing that the conservative way should be replaced by a new, socialist order.

The radical worldview is, hardly surprisingly, radically different and takes a different look at the meanings of the concepts of freedom and equality. Theoretical freedom of the press, for example, is seen as meaning little in a society where one group (or class) controls the mass media; and the freedom of one person to accumulate material goods implies that others have less freedom of choice because less is available to them. Rather than trying to strike a balance between liberty and equality, the radicals say that the two go hand in hand, if we have a true picture of what liberty really is – but even the radicals argue amongst themselves on this point.

Radicals seem to many people to be offering a basically negative approach. They want to overthrow present society, and are clear on what they *don't* like, but seem to offer little in the way of a constructive picture of the new socialist order they expect to emerge from the wreckage. Their counter to this argument is that since a democratic future world will be run according to the democratic wishes of its inhabitants, in response to the circumstances prevailing then, it is not up to us to try to dictate the lines on which it should be run. But in fact the seemingly negative goals of the radicals do provide a picture of the kind of future world they wish to see.

A world in which the divisions between manual and intellectual work are removed and in which work and leisure are integrated into one lifestyle is made possible,

197

in the radical worldview, by the technological level we have now reached. People who have the same needs should receive the same treatment from society regardless even of the amount or nature of work they do, and all should have equal opportunity to make the best use of their own abilities. But even in a high growth world this ideal will not be achieved for generations, as the radicals themselves realise.

Meanwhile, welfare and income levels should be equalised across countries – following, of course, individual revolutions within those countries to throw out the old capitalist ways. Aid from the rich to the poor nations becomes an automatic response to needs in an ideal socialist world, while within poor countries struggling to improve their lot there would be little difficulty in giving priority to real needs and avoiding any lust for luxury consumer goods, perhaps following the path pioneered by China and Cuba. And, most important of all in the short term, land and industry must be taken into public ownership since, according to the radical worldview, landowners and capitalists acting in their own self-interests are bound to squash any attempts at liberal reform.

As I have said before, none of the three worldviews is going to give us an accurate picture of how the world evolves, even restricting our interest to high growth/more equal futures. In a sense, the reformist view is the most realistic since it is not built upon either idealised theory but accepts a mixture of ideologies and the inevitable – but not necessarily insurmountable – problems of conflict that result. Certainly the prospects of worldwide revolution in the immediate future (the next 50–100 years) are remote, so that a mixture of ideologies must continue to rub shoulders in the real world at international level, even if individual nations choose the conservative or radical paths. On the other hand, the radicals might be thought more realistic in their assessment of the priorities needed to achieve a more equal world as quickly as possible. The conservatives, however, seem to be out on something of a limb in that they are not really advocating equality as a prime consideration in their calculations, so that the future worlds we are interested in, one of which we hope to live in ourselves, are not very likely to be reached by following a pure conservative route. Whichever route we do follow, though, we are now in a position to see what it would be like to work and live in a world of more equality achieved through rapid economic growth.

In the chosen examples which follow, I am in effect 'playing through' the relevant portions of some of the roles which have particular importance in the three kinds of future world.

Food and Health

We know now that the world can produce enough food to feed a greatly increased population, but also that the necessary growth in agricultural production will only be achieved by an effort to change the pattern of world growth. To achieve a successful high growth/more equal future world, the more optimistic estimates of possible change spelled out in Chapter Three will have to be turned into reality. This depends critically on a feedback with the health and vitality of the working population of poor farmers in the developing countries. Well-fed, healthy people are in a much better position not only to do the hard physical work required but to listen to new ideas – children are stunted both physically and mentally if they have insufficient food during growth, and many diseases induce a mental sluggishness as well as physical weakness. So it is natural to look at the questions of food and health together.

Improved nutrition improves resistance to disease, and general health can also be improved by improving water supplies, sewerage arrangements and so on as well as by the directly 'medical' methods of vaccination and use of antibiotics. In the rich countries, of course, other 'diseases' have become more common as infectious diseases have been brought under control and life-styles have changed.

Heart disease, chronic bronchitis and cancer are obvious examples. But it is not clear that such a development is inevitable, since some at least of these 'diseases of civilisation' are thought to be associated with 'unhealthy' diet – too much food rather than too little. Perhaps this is another case where the developing countries can learn from the mistakes of the rich.

Some causes of death and illness in the present rich countries can, however, be avoided. Pollution-causing illnesses such as silicosis among workers in some industries; airborne pollution causing widespread respiratory troubles; and the hazards of radiation. All these and others are, at least to some extent, within our control. Better understanding of these problems with appropriate regulation of pollution, less damaging diet, and leisure activities which encourage physical well-being are all ways in which the 'diseases of civilisation' can be tackled as more countries become rich enough to have cause to worry about them.

The quality of life, indeed, might well be improved directly as a result of a reduced availability of 'rich' foods in the present developed world. In a conservative or reformist world, basic foods might be expected to remain cheap in the present rich countries, with an increasing consumption (either directly or 'secondhand' through animals such as pigs) of such novel sources as processed vegetable protein or microbial protein grown on organic material. With the present poor becoming rich, more people would be chasing the limited amount of beef and other expensive foodstuffs available in the market place, forcing the price well into the luxury bracket. Such changes – especially if coupled with an increase in leisure time – might well encourage people to grow more of their own food either for variety or cheapness, even in the developed countries.

In a radical world, a similar path could be followed as a matter of principle, with a possibility that reduced meat consumption in one part of the world really could result in more grain being available for others. But either way the food of the future is likely to be nutritious without necessarily being exciting, with every incentive for the amateur gardeners and cooks to make their own contributions to improving the quality of life.

Work and Play

Taking an optimistic view – as I have, by and large, throughout this book – high growth in the developing world could be channelled into a quite different industrial pattern from that reached in the rich North. Instead of craftsmen and peasants moving into the cities to become factory workers, jobs could be redesigned by use of appropriate technology so that the work is taken out to the workers in the equivalent of cottage industries. The Chinese communes, encouraged to be as self-sufficient as possible even in the manufacture of their own machinery, provide one example of this kind of process, although it can be achieved in reformist as well as radical scenarios (but probably not in a 'pure' conservative world). Some reformists even argue that a similar de-centralising process can be achieved in the present industrial countries, providing scope for more interesting work and for job rotation, factors which could be contributors to an improved quality of life. In principle, many white-collar (paper pushing) jobs could now be transferred to people's homes, with improved telecommunications, which has the added advantage of saving energy as the need for workers to commute many miles into city centres is removed or reduced. Automation of tedious 'paperwork' by computers can also improve the enjoyment of work by white collar workers, just as machines have removed the burden of labour from many workers.*

Job mobility is already increasing in the rich North,

* The pessimists point also to the possibility of many paper pushers being put out of work by computers, leading to a depressed world peopled largely by the unemployed. This possibility is real, but can be avoided; I shall continue to develop here the theme of a *desirable* future world.

201

although largely for the wrong reasons – people are made redundant and forced to seek a different kind of work. If we continue to progress in a high growth world with rapid technical change, it will be necessary to adapt our society to the prospect of frequent retraining (by past standards) for many workers, and this too can improve the quality of life by providing variety instead of boredom. But one of the crucial problems now facing the industrialised rich countries is the question of unemployment, an issue related, in our weaving of future worlds, to the question of leisure activities in a different kind of society.

Because of technological change, increased automation and so on, a smaller proportion of the working population is employed directly in industry. Kahn estimates (*The Next 200 Years*, pp. 204–5) that by the year 2000 the US population may be 250 million of which 100 million or so will be employed. But only 25 million are likely to be needed in industry, the section of the economy directly geared to production. The rest – three quarters of the workforce – will be working for the government, teaching, in hospitals, or doing 'things judged worth doing for their own sake', such as the arts or fundamental scientific research. Many of these jobs 'will be meaningless in terms of product to society, though they may enjoy doing them: the jobs will merely be an accepted way of transferring income to such people', an alternative, if you like, to handouts from the State. And many other people, of course, will not be working at all, but will indeed be receiving their share of the national cake directly as what are now called welfare payments.

Our present society still has a horror of unemployment, a feeling that there is something sinful about wasting so many potential workers. Both politicians and union leaders were brought up during a period when such attitudes had some basis, but the situation is changing and employment in the conventional sense should not be seen as either necessary or desirable for every individual. If one eighth of the population is all we need to produce the goods we need, and if those workers have been freed from the burden of heavy manual work or boring paperwork, so that they have interesting and happy lives, why shouldn't the other seven eighths of the population also enjoy life? Such enjoyment could merely be entertainment, as in some science-fiction scenarios, although this is unlikely to be the case in a healthy 'best possible' future world. More likely, and probably a more welcome prospect in the immediate future, is a society based on a different *kind* of work, but still fulfilling any basic human need to participate in society. Some alternatives are discussed below in a section on post-industrial society, but one area likely to be affected by such developments is certainly education.

It is arguable, of course, whether or not teachers (along with doctors and administrators) should be counted as part of the productively employed workforce. True, if there were no teachers around tomorrow the world would keep going for a little while, but industrial socicty might begin to feel the strain if no teaching was done for, say, the next ten years. That said, however, I shall consider education in its conventional box as a 'service' rather than a 'productivc' industry.

Education offers scope both for more 'employment' – a way to distribute some of the national wealth – and a constructive occupation for people with ample leisure time. In addition, with growth continuing to produce a rapid rate of change, re-education would be increasingly necessary to enable people to keep in touch, either with developments affecting their work or simply for personal satisfaction.

Developments such as the 'Open University' in Britain may point a way for future developments in the industrialised countries; more dramatic impacts might occur in the developing world in the near future, with satellite transmissions used to aid teaching of practicalities related to the vitally necessary agricultural improvement, industrial development or use of local fuels to a scattered population in a region where conventional communications have never become fully effective.

The different worldviews differ less on how education might be carried out in the future – with computer assistance, video recordings and game-playing calculators – than on what would be taught. The conservatives stress the need for a return to 'traditional' discipline and use of the classroom to instil a sense of values appropriate to a capitalist world, while reformists would allow more experimentation in educational systems and radicals might see more need for the educational system to respond to needs expressed by those being taught instead of following any path laid down by teachers alone. Whatever your worldview, however, the crucial factor is to escape from the shackles of the work ethic which holds that unemployment or 'non-productive' employment is evil. If suitable provisions were being made to provide a decent way of life for all, it could then be a source of pride that in Britain, for example, $1\frac{1}{2}$ million less people are needed to grovel in coal mines or work at monotonous, production line jobs. Unemployment is neither good nor bad of itself, but depends upon the attitudes and responses of society and the individual to become either. Millions of civil servants, increased numbers of teachers and doctors and so on should not be seen as some kind of parasitic growth on the back of the productive workforce, or as an expensive luxury, but as a valid way of providing fulfilling occupations for many people who might otherwise be not just employed but, as a result, leading the lives of second class citizens, given the present structure of society.

Travel and Housing

Some of the most dramatic examples of the impact of changing technology in everyday life come from the area of transport – which includes the movement of ideas as well as of people and material goods. One jet airliner can now carry hundreds of passengers for thousands of miles between stops; one large tanker can carry 75 million gallons of gasoline – enough to keep even the most inefficient car running for quite a while – and communications via satellite have opened up the world to everyone with a TV – we are now blasé about watching live TV from another continent, and who now remembers the drama of watching the first steps of the first man on the Moon live via global TV coverage? Even the wildest science-fiction writers of the fifties never envisaged that possibility!

But to most people in industrialised Western society, travel is linked automatically with one phenomenon – the privately owned car. And the role of the car may change considerably in the future worlds of any of our three worldviews, even given high enough growth rates in the world as a whole to enable developing countries to catch up with the rich in many ways. There is no doubt that mass transport is more efficient in terms of energy use than individual transport, with a bus carrying thirty people using less fuel per passenger per mile than a car carrying only the driver and one passenger. A fully loaded Jumbo jet is equivalently more efficient than Concorde in these terms, and the shift from 50,000 tonne tankers up to 500,000 tonnes is claimed to save 70 per cent of transportation costs. Cars also use a lot of raw materials, although since American cars on average weigh twice as much as European cars there is plenty of room for improvement here. So a massive growth in the number of cars in the world would imply a massive consumption of energy and raw materials – but always keep in mind that it also implies a massive investment in industry to build the cars, and has implications for employment as well.

But just how much growth in car 'population' is likely? Early estimates that multi-car households would become common have been confounded by a levelling off in the growth curve – another 'S' shaped curve, it now seems. Since about 1960, growth rates for car numbers have declined in the OECD countries, which between them have

90 per cent of the present car population. If high growth took developing countries along a 'Western' path, we might expect to see a similar rise and levelling off in their car numbers, although the cars themselves might be more rugged and energy efficient than the luxurious gas-guzzlers of the present USA. This is not inevitable, though, since governments in the developing countries may prefer – especially in more egalitarian future worlds – to develop efficient public transport services which do away with the need for many private cars, which would be legislated against.

Starting from scratch, poor countries have a choice of several options. Canals, roads and railways all have their appeal in different circumstances and over different terrain; even the speculators who are trying to promote the use of airships may yet come into their own if they can find the right niche. Over long distances, with energy costs high and, perhaps, concern for the environment a prime consideration, the modern sailing ship designs which are promoted by present day 'environmentalists' might become a practical reality, and at a more mundane level pipelines can be used to move goods either pneumatically or mixed with water as a slurry.

I have already mentioned the possibility of replacing many journeys by telecommunications, a possibility which becomes increasingly attractive as communications systems continue to improve. But good communications do encourage some forms of travel, with people persuaded by what they see on TV to visit exotic places for holidays. And more leisure time may mean less travel to work but more travel to play. So while it is clear that the patterns of travel may change, it is far from clear that there will be less travel overall in the future. Apart from this, however, there are many significant communications developments still in the pipeline even if technological breakthroughs ceased tomorrow. Holography – true three dimensional imagery – and other laser applications, video cassette equipment, home computers linked to communications systems, or TV/telephone communications systems linked to centralised computers, wide use of optical fibres and other possibilities become very likely in a high growth world. With such growth, the ordinary citizen in the developed countries, and before too long in the whole world, is likely to have ready access to a vast amount of information and to a far-reaching communications system. There is a real possibility that the printed book or newspaper might be displaced by computer linked video systems, for example, and there is no way to do justice to the broad implications of such possibilities in the limited space available here. But watch out – for in my view it is in this area of communications that you and I will most quickly see the changes resulting from a high growth world.

Most readers of this book in the developed countries will see less obvious changes in housing. Where industrialisation has already taken place, the shift of population from the land to the cities has already occurred. But everywhere in the world the same pattern occurs with agriculture becoming more machine-oriented and industry in cities providing at least the hope of employment for many people. So shanty towns spring up around cities which don't expand fast enough to cope with these changes. With low growth, and/or inegalitarian policies, this picture might change little in the immediate future. In a high growth/ more equal world, however, basic facilities (water and sewerage) could be supplied to these townships and at least basic housing built to replace the shanties. Conservative scenarios might see this as an opportunity for multinationals to take Western technology and building practice into the developing world on a profitable basis; others might see local development of cement, steel and other industries to meet the demand for housing. Even an optimist, though, might feel a little gloomy about the prospects of such developments actually occurring, given the appalling track record of so many governments in this area in the recent past.

But at least in that case it is clear what needs to be

done. It is rather less obvious just what changes will be appropriate as the nature and size of the 'family' unit changes. Even looking only a hundred years ahead there are prospects of considerable shifts in housing needs, with perhaps more young people living alone and requiring small dwelling units, or more people preferring to live in some form of communal groups, requiring large dwelling units. Will the strength of, say, the present feminist movement grow and lead to a change in lifestyles which affects housing and domestic work (more families sharing cooking facilities, perhaps)? Will 'automation' move into the home to reduce further the need for 'housework' by providing disposable utensils (and furniture?) and mechanised cleaning? And above all, what will any of these developments mean for the quality of life?

The questions here are easier to find than the answers. But as we saw in the example of unemployment and kinds of employment, it is vitally important for those planning the development of our society to appreciate the kind of change that is going on, and to plan for the future, not for the past pattern of society as it was in their own youth. This, perhaps, is the place to look in broader terms at the way society might develop as we move beyond the growth phase of the 'S' shaped curve.

Post-Industrial Society

The conventional economic wisdom about what happens to society after the phase of industrialisation that the developed world has now all but gone through is that it moves into a 'service' economy. Figure 28 shows the basis for this argument, known as Engel's Law. At any given time in any one country, it seems to follow that the more money an individual has then the greater proportion is spent on 'services' compared with food or durable goods. A service is, by definition, something which is not durable and in the words of another pioneering economist, Adam Smith, the products of servants '. . . perish instantly in

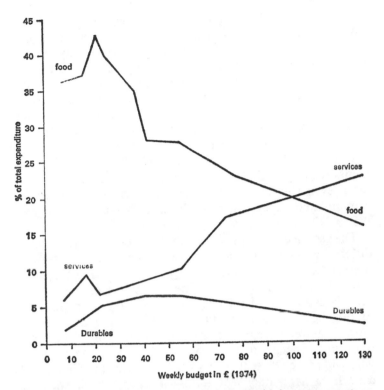

Figure 28: How the proportion of household income spent on different items varied between rich and poor households in the UK in 1954 – an example of 'Engel's Law'. Data from the *Family Expenditure Survey*, cited by J. I. Gershuny, 'The Self-Service Economy', *New Universities Quarterly*, vol. 32, p 50, 1977

the instant of their production'. In this sense, the performance of a play by actors is a service not a 'good', as would be the duties of those we think of traditionally as servants, the chauffeur driving a car, for example. Once the journey is over that service is gone for ever, whereas the car itself, although 'used' to some extent, is still there to use again, and eventually to be recycled.

Engel's Law makes good common sense. Obviously one

person (or one household) can only eat so much food, however much income is available, so the percentage of income spent on food falls as income rises way beyond that sufficient to provide the essentials of life. Even durable goods provide limited scope for expenditure for any sane 'consumer', however rich. But services are another matter. Once you own a private jet there is little point in buying another (and even less in buying a third). But there is every point in hiring a pilot, crew, personal secretary and assistants to keep you informed of business developments while you flit about the globe. At a more mundane level, even the cost of a Rolls Royce is far less than the accumulated expenditure on the wages of a chauffeur over the car's lifetime.

The service economy concept takes this picture of differences in society at any one time and applies it to predict how society may change as everyone gets richer through economic growth. One day, it seems, we will all have 'servants', and we will all spend a bigger proportion of our household budgets on services than on food or durable goods – or on both put together. This is a slightly worrying concept in many ways. If we all have servants, who will serve the servants – who will drive the chauffeur's car? It rather smacks of taking in each other's washing, and we can't all be actors or who will watch the plays? Indeed, the evidence is that this simplistic attempt to describe post-industrial society doesn't really work. In the UK, between 1954 (the date of the data of Figure 28) and 1974 economic growth certainly did bring more disposable income to just about everybody. Yet the same source as that of Figure 28 shows the percentage of income spent on services over that twenty-year period almost steady at 9·5 per cent. Since the price of services rose twice as fast as that of goods over the same period, the real proportion of consumption of services must have been half that of twenty years before by 1974. So where does the simple theory break down?

Economist Dr Jay Gershuny has presented an alternative

view of post-industrial society, as what he calls the 'self-service' economy. In a nutshell, what happens is not that we take in each other's washing, but that we each buy a washing machine; we don't chauffeur each other, but each drive our own car; we don't even go to the theatre very much, but watch TV at home. And all of this involves expenditure on goods, not on services, counter to conventional economic wisdom. Whether you should class manufacture of TVs as production of goods or supply of services, then, might become a debating point for economists. Much more interesting in the context of possible future worlds is the question of how far this trend can go – not least since it is closely linked with some forms of decentralisation and even more closely with growing amounts of 'leisure time', 'unemployment' or whatever you choose to call it.

Increasingly, households buy goods which are installed in the home and used *together with the household members' own labour* to produce items of final consumption. Do it yourself furniture, building up hi-fi equipment, already even build it yourself computers and calculators; home education through TV (Open University) and correspondence courses; perhaps even some aspects of medicine as people are encouraged to have babies at home, for example, with minimum trained supervision, calling in a doctor only in emergency (and quickly, with efficient transport and communications systems). It may look more like science fiction than economic forecasting, but the change that has virtually replaced the railway as a system of passenger transport by the private car is no less striking. Gershuny suggests that this implies a need to change the direction of policy aims in the industrialised world, the rich North, not trying to maintain the status quo of the old ways and boost conventional employment, but encouraging progress towards the self-service economy. More goods might be provided in component form ready for home assembly; domestic 'production' could become a satisfying alternative to formal employment, and people in general would be much more free to 'do their own thing'. The trend seems to be already happening anyway; the argument is that it is better to bend with the breeze than to try to fight against it.

All this takes us beyond the immediate scope of the bulk of this book. The problems of post-industrial society are, if not quite those of having too much of a good thing, less pressing than the problems of bringing adequate food and as much as possible of 'the good life' to the inhabitants of the poorer nations of the world. It's good to have some idea of what might lie ahead, even in speculative form, but what really matters is whether it is possible for the developing countries to make enough progress quickly enough to bring equality to the world and leave everyone to face the problems of post-industrial society together. So now it's own up time – can it really be done, or is it all pie in the sky?

Is It Possible?

The optimistic tone I have tried to adopt should not be allowed to disguise the important distinction between *possible* future worlds and *probable* developments from the present day situation. The path to a desirable future is a narrow and difficult one, and it is not necessary to be a dyed in the wool pessimist to see other alternatives such as high or low growth unequal worlds, with or without major wars, as more easy to 'achieve', simply by letting things drift along without any direct effort to improve them. What matters, from the optimistic point of view, is to establish that a better alternative can be achieved, and to establish this as a goal towards which efforts can be directed. By shooting for the stars, we might at least hit the Moon; if we don't bother with any direct effort to improve our lot things will certainly get worse.

But progress towards a desirable future world will not be easy, if it is achieved at all. I have argued that growth is an essential ingredient of this progress, but this is not to say that all growth is good. Undirected growth is more likely to produce inequality, since the present rich start out from a favourable position, and inegalitarian high growth worlds are unpleasant for most of the people living in them – the teeming poor. It is this danger which makes a low growth, egalitarian world attractive to many radicals as a short-term objective, with high growth following after equality has been achieved.

Such worlds also have their problems, since where, in a low growth future, can the material and economic resources to improve the lot of the poor come from? The fact that the analysis reported here shows that there are no insurmountable physical reasons why the material problems of the world (food, raw materials, even energy supply) cannot be solved in a matter of decades rather than centuries, is, surely, sufficient reason to make the attempt to achieve this goal.

This discovery that the carrying capacity of Planet Earth is not, after all, being exceeded – so new and surprising a discovery to so many people – is doubly important because if we can remove the neo-Malthusian arguments from the centre of the stage to the wings where they belong, it becomes much easier to focus attention on the real major problems of global survival into the era of the good life. The social, political, and especially military problems now facing us are far more important than the arguments about physical limits; if the energy directed by so many people into the limits debate could be diverted into these problem areas, then perhaps there would be a better chance of survival. So it is appropriate, perhaps, to close this particular book with an optimistic reminder of how greater equality can be achieved with the aid of growth, while acknowledging that there are many other less pleasant future worlds lying in wait for us if we try to choose the soft options.

In agriculture, even if the world's population grows much more rapidly than anyone expects over the next 50–100 years, physical limits on the production of food will not be a major problem. The problem which must be solved is the distribution of available food to those who need it, which with the present world system means that the poor must be provided with the means to earn money with which to buy food. In a nutshell, the problem is one of income distribution – a social and political problem, not one of physical shortages.

Transition into a world where the present poor countries can cease their dependence on imported food and become self-sufficient will not be easy, and will require basic changes in political, economic and technological habits, as well as increased inputs of both materials and energy. Partly through this link, the energy supply of our desirable future world is the key sector.

Within the energy sector, the key *issue* is whether the drive for equality can be fuelled without recourse to widespread use of nuclear fission power, with resulting dangers from radioactive wastes and from the increased possibilities of nuclear war, sabotage or major accidents. Because of the dangers of this path, it seems essential to make a major effort to develop, in particular, the large available stocks of coal, with some contribution being made from geothermal, solar and other less conventional sources. At the same time, so much energy can be saved by improving the efficiency levels that were until recently acceptable that there is overall a real prospect of avoiding the need for a fission-based energy future. With the right policies, the difficult road to a desirable future can be followed; and those policies need to be implemented now, before all our eggs are committed to the nuclear basket.

The supply and use of raw materials may be the least of our three main problems, but that doesn't mean the problems involved are insignificant. Once again, as with food, the important issues are not those of physical limits, but of technology, politics, economics, and preservation of the environment. The use of particular materials for particular tasks will certainly change through substitution, and some materials may be in short supply. But something will always be available to fit any particular need, if the economic resources are available for development. This, of course, is another argument for economic growth, without which we might be unable to tap adequately the almost inexhaustible (by human standards) storehouse of materials in the Earth's crust.

In one important way the role of raw materials in the future is different from the problems of world food. Whereas with food appropriate redistribution of *existing* supplies would prevent malnutrition, with materials increased production is an integral part of any high growth route to equality. Industries are built from raw materials, and industrial growth in the poor countries is a vital ingredient of our preferred future world. Increasing production of raw materials once again needs more energy – quite apart from the energy needed by growing industries. Other future worlds might look more attractive in terms of energy requirements, or material needs, or even food distribution, if these aspects of the whole are looked at in isolation. But this is a dangerously misleading way of looking at world problems; only by looking at the whole can we really find the best alternative, with the least prospect of conflict and misery in the longer term. The words of Professor Marie Jahoda, the head of the SPRU forecasting team, provide an appropriate note on which to end this study of the physical background to the choice of future worlds now facing us, and to bear in mind as attention turns to the vital political and economic problems of making the best use of the resources we have:

As long as there is some rational basis for thinking that the world's population could have enough to eat, an adequate material base, a satisfactory quality of life and less fear of war, it would be irresponsible to strive for less.

A Guide to the Great Futures Debate

Many books are now available which present particular viewpoints in the futures debate, and all too many of them present one point of view as 'the' path to the future. This makes it particularly important in this highly controversial area of study to read as widely as possible, and I would be unhappy if anyone read just this book, for example, and took my word as gospel truth. The following list of related works is not exhaustive, but combines the texts which I have found particularly useful and those which are particularly important either for historical reasons or for their content.

General 'futures' books

Pride of place must go to: *World Futures: The Great Debate*, ed. C. Freeman and M. Jahoda (Martin Robertson, London, 1978; Universe, New York, 1978). This is the most comprehensive overview of the entire futures debate yet available, and provides much of the background to the present book.

Any book which made a major impact on the popular consciousness and brought awareness of the need to think about the future to millions of people deserves attention, so read: *The Limits to Growth*, D. H. Meadows, D. L. Meadows, J. Randers and W. W. Behrens (Pan, London, 1972).

216

But don't read this in isolation – compare it with: *Thinking About the Future*, ed. H. S. D. Cole, C. Freeman, M. Jahoda and K. L. R. Pavitt (Sussex Univ. Press/ Chatto & Windus, London, 1973).

Of the rest, some of the most readable (but to be taken generally with a pinch of salt and compared with one another) are:

Michael Allaby, *Inventing Tomorrow* (Abacus, London, 1977) (Particularly relevant to the UK and Europe).

S. Encel, P. K. Marstrand and W. Page (eds.), *The Art of Anticipation* (Martin Robertson, London, 1975; Universe, New York, 1975).

Robert Heilbroner, *An Enquiry into the Human Prospect* (Calder & Boyars, London, 1975) (Take this one with TWO pinches of salt!).

Herman Kahn, William Brown and Leon Martel, *The Next 200 Years* (Associated Business Programmes, London, 1976 and Abacus, London, 1978).

M. Mesarovic and E. Pestel, *Mankind at the Turning Point* (Hutchinson, London, 1975).

E. F. Schumacher, *Small is Beautiful* (Abacus, London, 1974).

T. Whiston, (ed.) *Uses and Abuses of Forecasting* (Macmillan, London, 1979).

Books related to specific topics discussed here

P. L. Cook and A. J. Surrey, *Energy Policy: Strategies for Uncertainty* (Martin Robertson, London, 1977).

Antony Flew (ed.), *Malthus: An Essay on the Principle of Population* (Pelican, London, 1970).

Susan George, *How the Other Half Dies* (Pelican, London, 1976).

J. I. Gershuny, *After Industrial Society* (Macmillan, London, 1978).

John Gribbin, *Our Changing Planet* (Wildwood, London, 1977; Crowell, New York, 1977; Abacus, London, 1979).

John Gribbin, *The Climatic Threat* (Fontana, London, 1978; Scribner's, New York, 1979; US title *What's Wrong with our Weather?*).

John Gribbin, (ed.), *Climatic Change* (Cambridge Univ. Press, London & New York, 1978).

K. Griffin, *The Political Economy of Agrarian Change* (Macmillan, London, 1974).

Herman Kahn, *Thinking About the Unthinkable* (Weidenfeld, London, 1962).

NAS, *Energy and Climate* (National Academy of Sciences, Washington, DC, 1977).

Gerard K. O'Neill, *The High Frontier* (Cape, London, 1977).

Index

219

222

223